INTERGOVERNMENTAL RELATIONS AND ECONOMIC MANAGEMENT IN CHINA

STUDIES ON THE CHINESE ECONOMY

General Editors: Peter Nolan, Lecturer in Economics and Politics, University of Cambridge, and Fellow and Director of Studies in Economics, Jesus College, Cambridge, England; and Dong Fureng, Professor, Chinese Academy of Social Sciences, Beijing, China

This series analyses issues in China's current economic development, and sheds light upon that process by examining China's economic history. It contains a wide range of books on the Chinese economy past and present, and includes not only studies written by leading Western authorities, but also translations of the most important works on the Chinese economy produced within China. It intends to make a major contribution towards understanding this immensely important part of the world economy.

Published titles include:

Derong Chen
CHINESE FIRMS BETWEEN HIERACHY AND MARKET

Dong Fureng
INDUSTRIALIZATION AND CHINA'S RURAL MODERNIZATION

Du Runsheng (*edited by Thomas R. Gottschang*)
REFORM AND DEVELOPMENT IN RURAL CHINA

Qimiao Fan and Peter Nolan (*editors*)
CHINA'S ECONOMIC REFORMS

Christopher Findlay, Andrew Watson and Harry X. Wu (*editors*)
RURAL ENTERPRISES IN CHINA

Michael Korzec
LABOUR AND THE FAILURE OF REFORM IN CHINA

Yuming Sheng
INTERSECTORAL RESOURCE FLOWS AND CHINA'S ECONOMIC DEVELOPMENT

Hong Wang
CHINA'S EXPORTS SINCE 1979

Tim Wright (*editor*)
THE CHINESE ECONOMY IN THE EARLY TWENTIETH CENTURY

Intergovernmental Relations and Economic Management in China

Jun Ma

Public Policy Specialist
The World Bank
Washington, DC

First published by
MACMILLAN PRESS LTD
Houndmills, Basingstoke, Hampshire RG21 6XS
and London
Companies and representatives
throughout the world

ISBN 0–333–66007–2

A catalogue record for this book is available
from the British Library.

This book is printed on paper suitable for recycling and made from fully managed and sustained forest sources.

Transferred to digital printing 2000

Printed in Great Britain by
Antony Rowe Ltd, Chippenham, Wiltshire

To my wife and my parents

Contents

List of Tables

List of Figures

List of Appendix Tables

Acknowledgements

This book is largely based on my PhD dissertation at Georgetown University, entitled 'Four Essays on Intergovernmental Fiscal Relations, Intergovernmental Monetary Relations, and Policy Changes in China' and my research report on 'Macroeconomic Management and Intergovernmental Relations in China', prepared for the Public Economics Division of the World Bank. During the course of these studies, I benefited from a vast number of people in the US and in China. I am particularly indebted to Professor Carol Rogers, who read every version of my dissertation and provided me with invaluable advice on issues ranging from basic theory to modeling techniques. Special thanks are also due to Professors William Hoyt and Serge Moresi for many helpful suggestions and corrections. Professors Richard Sweeney, David Walker, Marius Schwartz, and Stuart Brown also shared with me their valuable insights.

On many policy issues that inspired the theoretical work, discussions with economists that I know from my work at the World Bank and the International Monetary Fund are especially useful. These colleagues are, among others, Anwar Shah, Rajiv Lall, Hyung-ki Kim, Anjali Kumar, Heng-fu Zou, Bert Hofman, Peter Harrold, Albert Keidel, Kangbin Zheng, Michael Bell, and Hoe Ee Khor. Comments from conference and seminar participants at Georgetown University, Rutgers University, University of Minnesota, University of Chicago, the annual meetings of the Chinese Economist Society (North America), the monthly seminars of the Chinese Professional Forum (Washington, DC), the 1994 and 1995 annual meetings of the American Economist Association, and the 1995 annual meeting of the Association for Asian Studies also helped me develop and sharpen my arguments.

I also thank my friends in China, from whom I obtained much raw material and was informed about the current thinking of Chinese researchers and policy makers. These people include Lu Baifu from the Development Research Center of the State Council, Wong Weixing, Zhao Hong, and Zhu Guangyao of the Ministry of Finance, Jiang Wangjin from the People's Bank of China, Guo Shuqing and Liu He from the State Planning Commission, Shen Hong from the Chinese Academy of Social Science, Wang Zhan from Shanghai Economic Research Center, Lu Deming from Fudan University Economic Research Center, Yi Zhengqiu from Guangdong System Reform Commission, and Xu Jingan from Shenzhen System Reform Commission.

I enjoyed working with the editors of the series of Studies on the Chinese Economy and representatives of Macmillan Press in producing this book. Special thanks to Professor Peter Nolan, who carefully reviewed the manuscript and Mr Ray Addicott, Ms Giovanna Davitti, who oversaw the copy-editing and production process.

Portions of some chapters in this book appear in my articles in several journals. These include 'The Reform of China's Intergovernmental Fiscal Relations' in *Asian Economic Journal* (vol. 9, no. 3, 1995, pp. 205–32), 'Modelling Central–Local Fiscal Relations in China' in *China Economic Review* (vol. 6, no. 1, pp. 105–37), 'Monetary Management and Intergovernmental Relations in China' in *World Development* (vol. 24, no. 1, 1996, pp. 145–53), 'China: Central Government Credibility and Economic Overheating' in *Economic Systems* (vol. 19, no. 3, 1995, pp. 237–61), and 'Defining the Limits of Local Government Power in China' in *The Journal of Contemporary China* (no. 10, 1995, pp. 3–22). My thanks to East Asian Economic Association, JAI Press Inc., Elsevier Science Ltd, Psysica-Verlag, and Center for Modern China for permissions to use these materials.

I owe a special debt to my wife and my parents. My wife, Ching Chung, commented on almost every chapter of the book. Although she is not an economist, her comments showed a great deal of insight. Her constant encouragement and support, as well as those from my parents, have been one of the most valuable inputs to this book.

Finally, I shall claim that I am the only person responsible for any remaining errors of this book. The views expressed in the book are mine and should not be attributed to the World Bank.

List of Abbreviations

ABC: Agricultural Bank of China
BOC: Bank of China
BOCOM: Bank of Communications
CITIC: China International Trust and Investment Corporation
COE: collectively owned enterprise
COV: coefficient of variation
FBIS: Federal Bureau of Information Service
FERT: Foreign Economic Relations and Trade
FTC: Foreign Trade Corporation
GDP: gross domestic products
GNP: gross national products
ICBC: Industrial and Commercial Bank of China
IMF: International Monetary Fund
MOF: Ministry of Finance
MOFERT: Ministry of Foreign Economic Relations and Trade
NBFI: non-bank financial institutions
NPC: National People's Congress
OLS: ordinary least square
PBC: People's Bank of China
PCBC: Peoples Construction Bank of China
PICC: Peoples Insurance Company of China
SAIC: State Administration of Industry and Commerce
SD: standard deviation
SEZ: special economic zone
SOE: state-owned enterprise
SINOPEC: China Petrochemical Corporation
VAT: value added tax

Introduction

China has a population of 1.2 billion and covers an area of 9.6 million square kilometers. Administratively, the country is divided into 30 provincial-level localities, including 22 provinces, three municipalities directly under the central government, and five autonomous regions (unless otherwise stated, we hereafter refer to these provincial-level localities as provinces).[1] The average population size of a province is 40 million, larger than most states and provinces in other large countries. Under the provinces, there are three administrative levels. They include 335 prefectures and municipalities at the prefectural level; 2166 counties, autonomous counties and cities at the county level, and townships and towns and city districts.[2] China's vast population and territory, as well as its geographical and economic diversity, suggest the great importance of policy coordination between the central and subnational governments.

As the economic reform in China has evolved over the past 15 years, most of the features associated with the traditional and centralized planning system have been replaced by new and more complex features of a decentralized system. The economic decision-making system has been greatly decentralized from the center to the localities, which now have substantial power in areas such as revenue collection, government expenditure, credit allocation, investment project approval, price and wage control, foreign trade management and industrial policy formation. Inefficiency due to the rigidity of a central planning system is no longer a significant issue. Instead, the danger of the central government losing control over revenue collection and credit expansion has become apparent. In contrast to the situation in the pre-reform era, increasing fiscal deficits, high inflation, and frequent economic cycles are now the main concerns of the policy makers.

At an abstract level, the old system dictates that the local governments can play only limited, usually passive, roles in the economy, while the decentralized system offers them a greater variety of choice variables and, in many cases, allows them to act as the 'leader' in their interaction with the central government. In other words, the new system can be described as yielding more complex and interesting principal–agent problems. These new features call for serious research which can explain the functioning of the decentralized system. The existing literature has largely ignored this issue, and a few relevant studies remain at the institutional stage. Moreover, the traditional approaches to forecasting the impact of government policy –

which consider the economic agents (local governments, enterprises, or consumers) as passive 'policy takers' – are no longer appropriate in the Chinese context.

This book examines how central–local economic relations have affected and will affect the central government's ability to control key macroeconomic variables and the functioning of the market. Its implications have both positive and normative elements. From a positive point of view, the institutional discussions as well as the theoretical models and simulations describe the behavior of the central and local governments under alternative institutional environments. In particular, they explain how China's macroeconomic performance could be an equilibrium of the interaction between levels of the government and explore the welfare implications of the results. From the normative point of view, the study suggests how to establish desirable rules of games in China's political and economic institutions through appropriate reforms and to achieve outcomes that are closer to the first best. Specifically, it recommends to the Chinese authorities, with reference to the experiences of other countries, the principles of division of powers between different levels of government and the specific measures needed for improving the central government's ability to stabilize the economy, to redistribute income, and to promote market competition.

This study employs analytical tools of theoretical modeling and simulation. The modeling part of this study will be based on a principal–agent set-up with one principal (the central government) and possibly more than one agent (the local government) that describes the mechanisms of China's central–local fiscal and monetary interactions. Chapter 2 considers a game between the center (the principal), whose choice variable is the revenue-sharing schedule with the localities, and the localities (the agents), whose choice variable is the efforts of revenue collection. Chapter 5 develops a central–local monetary game between the central bank and local banks, the central bank's choice variable being the credit plan and the local banks' choice variable being credit allocation across projects. Both models have the feature that the central government/bank and local governments/banks are treated as rational agents, who maximize their objectives subject to relevant constraints. The theoretical part of the work is mainly to identify the reasonable specification of each agent's objective function and the constraints that he/she faces. In both models, the parties involved have different objectives, some of which are the results of cross-region externalities. While the agents (localities) differ from the principal (the center) in objective, the latter has no simple authority to enforce directly its desired plan at the agent level. This approach has been novel in the literature on China's central–local relations.

The key questions discussed in this book are as follows.

(1) On fiscal management: how do the central–local fiscal relations affect the central government's ability to achieve the goals of stabilization and equalization through fiscal policies? How do the central–local fiscal relations affect local governments' effort to collect revenues and incentive to lobby for central government transfers? What are the alternative arrangements that can improve the central government's ability to use tax policy, increase local tax efforts, and control excessive expenditure expansion?
(2) On monetary management: how do the central–local monetary relations affect the central bank's ability to control money supply? What is the role of each institutional element of the central–local monetary relations (for example, local government–local bank relations, policy lending and commercial lending, central bank independence, and so on) in creating inflationary pressures?
(3) On market development: how does the *de facto* or *de jure* division of regulatory powers between levels of government affect the local governments' incentive and ability to restrict interregional trade, interregional capital flow, industrial reorganization, and market competition?

It is argued in this book that as for fiscal policy, the central–local fiscal contract system adopted in the early 1980s has left the central government with very limited policy instruments. The localities have obtained *de facto* control over effective tax rates and tax bases, while the lack of coordination between the decentralization of revenue collection and the decentralization of expenditure responsibility has restricted the center's flexibility in using expenditure policy. As the central government loses the ability to control tax bases and tax rates, it resorts to various *ad hoc* measures to adjust the central–local revenue division. These *ad hoc* measures have created serious incentive problems which have led to reduced local tax efforts and consequently weakened the center's fiscal capacity. Moreover, the lack of a rule-based intergovernmental transfer mechanism has caused excessive lobbying by the local governments and excessive expenditure expansion of the central government. As for monetary policy, it is argued that the central bank has not been able to control the money supply effectively as the localities could react strategically to the central bank's credit policy. The regions have found ways to force the central bank to revise the credit ceiling upwards, thereby creating inflation. As for market development, the book points out that under the decentralized system local governments have tended to abuse their vaguely defined administrative and regulatory powers in order to protect local economies. Such local protectionist behaviors have severely restricted the role of the market in resource allocation. While some aspects of the above issues are being addressed by the 1994 reform package, solutions to some other aspects are yet to be developed. Although the detailed design of a new framework of central–local relations is beyond the scope of this book, we will attempt to highlight selected elements of the needed reforms in the future.

The rest of this book is organized as follows. Chapter 1 provides an overview of China's intergovernmental fiscal relations and discusses their impact on the effectiveness of central government's fiscal policy. The 1994 fiscal reform plan is described and assessed, and areas that require further institutional changes are discussed. Chapter 2 presents a game theoretic model of China's central–local fiscal relations, focusing on the impact of the central government's lack of credibility in enforcing pre-announced revenue-sharing formulae on local governments' tax efforts. The model explains, at least partially, why the revenue : GNP ratio has declined rapidly over the past decade and a half. Chapter 3 looks at the expenditure side of the central–local fiscal relations. It shows how the Chinese system characterized by soft budget constraint – regions can effectively lobby for central government transfers after the Congress has approved the budget – caused excessive expenditure expansion. Chapters 4 and 5 turn to monetary management. Chapters 4 covers a review of China's financial system and a discussion on problems of the central–local monetary relations that have led to the central bank's failure to control money supply. This chapter concludes with an assessment of the 1994 monetary reform. Chapter 5 formalizes the discussion on central–local monetary relations in a game theoretic model and identifies the role of each institutional factor in the creation of excessive money supply. The model's finding suggests that the central bank's inability to commit to a pre-announced credit allocation plan is the key to understanding China's inflation. Chapter 6 deals with the division of central–local regulatory power and its impact on market development. Various problems arising from local governments' anti-competition behaviors are identified and a new regulatory framework that aims at limiting local government power abuses is proposed.

1 Intergovernmental Fiscal Relations and Macroeconomic Management

INTRODUCTION

The evolution of central–local fiscal relations has been the most important element of China's decentralization process over the past 15 years. It is widely agreed that fiscal decentralization has provided local governments with enormous incentives and fiscal powers to develop local economies and regional competition under the decentralized system has encouraged policy innovations at the local level. This chapter, however, will focus on the impact of fiscal decentralization on the central government's ability to achieve macroeconomic objectives, an area that has been paid insufficient attention but that is of extreme importance.

This chapter will argue that, under the fiscal contact system introduced in the early 1980s, the localities obtained *de facto* control over effective tax rates and tax bases and, thus, seriously undermined the central government's ability to use fiscal policy instruments to achieve goals such as macroeconomic stabilization and regional equity. The tax-assignment system adopted in 1994 addressed some of the major problems under the fiscal contract system, mainly through the centralization of revenues sources and tax administration. Nevertheless, fiscal reform in China is far from complete. A number of issues, including the stability of central–local fiscal relations, ownership-based tax assignment, division of expenditure responsibilities, and design of intergovernmental transfers, are still unaddressed and may continue to impact negatively on the functioning of the central government's fiscal policy.

The rest of this chapter is organized as follows. The following two sections provide an overview of the evolution of China's central–local revenue-sharing relations and the central–local division of expenditure responsibilities. The third section sketches the analytical framework that provides guidelines for assessing the performance of China's fiscal management system. The fourth section looks at how, under the fiscal contract system, the localities obtained control over effective tax rates and tax bases, leaving the center with very limited tax and expenditure policy instruments. In the fifth section the book examines the impact of the contract system on the center's ability to achieve goals such as macroeconomic stabilization and regional equalization. The 1994 fiscal reform and its implications are discussed in the the sixth and the seventh sections. The last section concludes.

REVENUE SHARING BETWEEN THE CENTRAL AND LOCAL GOVERNMENTS

Before 1980, the fiscal system was characterized by centralized revenue collection and centralized fiscal transfers, that is, all taxes and profits were remitted to the central government and then transferred back to the provinces according to expenditure needs approved by the center. Since the early 1980s, the central–provincial fiscal relation has gone through three major phases.[1] In 1980, the highly centralized system was changed into a revenue-sharing system in which the central and provincial governments each began to 'eat in separate kitchen'. This system was named as the contract responsibility system, which emulated a similar system adopted in the rural reform in the late 1970s.[2] For 15 provinces and autonomous regions, revenues were divided into three types: central-fixed revenues (those revenues and taxes accrued to the center), local-fixed revenues (those accrued to the localities), and shared revenues (those revenues shared between the center and localities according to some agreed rules).[3] During the period 1980–4, approximately 80 per cent of the shared revenues were remitted to the central government and 20 per cent retained by the local governments. Other provinces, cities, and autonomous regions adopted a variety of revenue-sharing methods, including fixed amount remittances to the center and fixed amount subsidies by the center, or continued to operate under the old system. All the revenues, except a few minor ones in the central-fixed revenues (such as customs duties), were collected by the local finance bureaus. The bases and rates of all the taxes, shared or fixed, were determined by the central government.

Although the 1980–4 fiscal relation initiated some localities' enthusiasm in collecting revenues, the uniform sharing formula created undesirable surpluses in wealthy provinces and deficits in poor provinces. In 1985, the State Council redesigned the revenue-sharing arrangement by setting varying schedules based on localities' budget balances in the previous years.[4] The principle was as follows: a province whose local-fixed revenue exceeded its expenditure should remit a certain portion of revenue to the center, a province whose local-fixed revenue fell short of its expenditure should be given a proportion of the shared revenue, and if the total of the local-fixed revenue and the shared revenue still could not cover a province's expenditure, the center should grant a fixed subsidy. Under this system, the central government allowed financially weak regions to retain more revenues or receive more subsidies, while continuing to maintain a tight grip over rich regions, such as Shanghai, Beijing, Tianjin, Liaoning, Jiangsu, and Zhejiang, that were the most important sources of the central revenue. This system controlled the horizontal inequality of revenue distribution, but it also reduced rich regions' incentive to expand their tax bases. During the period 1985–8, revenues collected by the rich regions generally grew more slowly than the national average. To mitigate this effect, in 1988 the State Council decided to adopt a new system in which six types of central–provincial revenue-sharing methods, each applied to a number of provinces, were introduced.[5] These six types are as follows.

(1) Contracted revenue increase (adopted by ten localities): based on 1987 revenue figures, localities retain a specific proportion ranging from 28 to 80 per cent of any revenue that is within a certain percentage growth (ranging from 3.5 to 6.5 per cent; any revenue in excess of 6.5 per cent growth is retained entirely by the localities).
(2) Uniform rate of retention (adopted by three localities): a fixed share retained by localities (46.5 to 87.6 per cent depending on the region).
(3) Variable rate of remittance (adopted by three localities): a certain proportion is retained up to a quota and then a different (usually higher) proportion of revenue is retained in excess of the quota.
(4) Increased amount contracted to go to the center (adopted by three localities): a specific amount is contracted to go to the center in the initial year; in subsequent years the amount is to increase at a contracted rate from 7 to 9 per cent.
(5) Fixed quota (three localities): a specific amount is contracted to be transferred to the center with no annual adjustment for high growth.
(6) Contracted transfers to provinces (15 localities): a specific amount is contracted to be transferred from the center to the localities.[6]

This 1988 fiscal contract system further increased the revenue share retained by the localities, in particular those major contributors to the central government's revenue. However, the contracts were not strictly adhered to and were revised repeatedly for some regions. In 1991, when the 1988–90 system was supposed to expire, the central government found itself unable to

TABLE 1.1 *Ratio of government revenue to GNP, 1978–93 (billion yuan)*

Year	*Government revenue*		*Government expenditure*		*revenue/ GNP* (per cent)*	*Deficit*†	*Deficit/ GNP (per cent)*
	Chinese Definition	*Standard Definition*	*Chinese Definition*	*Standard Definition*			
1978	112.1	124.8	111.1	123.8	34.8	–1.0	–0.3
1979	110.3	126.7	127.4	147.4	31.7	20.7	5.2
1980	108.5	130.1	121.3	147.1	29.1	17.0	3.8
1981	109.0	130.2	111.5	140.1	27.3	9.9	2.1
1982	112.4	140.9	115.3	152.3	27.1	11.3	2.2
1983	124.9	160.7	129.3	173.0	27.7	12.3	2.1
1984	150.2	184.3	154.6	196.5	26.5	12.2	1.8
1985	186.6	229.7	184.5	236.5	26.8	6.8	0.8
1986	226.0	244.6	233.1	265.6	25.2	20.9	2.2
1987	236.9	257.5	244.9	282.5	22.8	25.0	2.2
1988	262.8	280.4	270.7	315.3	19.9	34.9	2.5
1989	294.8	326.4	304.0	363.8	20.4	37.4	2.3
1990	331.3	351.5	345.2	403.1	19.9	51.6	2.9
1991	361.1	365.9	381.4	432.3	18.1	66.4	3.3
1992	415.3	392.9	439.0	483.4	16.1	90.5	3.7
1993	508.8	408.9	528.7	569.8	13.0	160.9	5.1

* Revenues using standard definitions.
†Expenditures minus revenues using standard definitions.

SOURCE McKinnon (1993) and State Statistical Bureau (1992a, 1994).
NOTE: the standard definition of revenue means subtracting borrowing from the Chinese definition and adding to the Chinese definition subsidies that were counted as negative revenue. The standard definition of expenditure means adding to the Chinese definition subsidies that were considered negative subsidies.

negotiate satisfactory replacements. As a result, the 1988–90 system was extended until the end of 1993, with limited modifications on revenue-sharing ratios and quotas.

Since 1978, both the total government revenue as a percentage of GNP and the share of central government revenue in total government revenue have declined rapidly. From 1978 to 1993, the ratio of total government revenue to GNP declined from 35 to approximately 13 per cent (see Table 1.1); the ratio of central government revenue (after revenue sharing) to total government revenue declined from 51 to 37 per cent (see Table 1.2).[7] Since the late

TABLE 1.2 *Change of central–local relative fiscal status, 1978–93**

Year	*Revenue*[†]					*Expenditure*				
	Total (*billion yuan*)	*Central*	*Local*	*Central* (*%*)	*Local* (*%*)	*Total* (*billion yuan*)	*Central*	*Local*	*Central* (*%*)	*Local* (*%*)
1978	112.1					111.1	51.0	60.1		
1979	110.3	15.8	94.6	14.3	85.7	127.4	64.7	62.7	50.8	49.2
1980	108.5	17.4	91.2	16.0	84.0	121.3	65.1	56.2	53.7	46.3
1981	108.9	22.5	86.5	20.6	79.4	111.5	60.2	51.3	54.0	46.0
1982	112.4	25.8	86.5	23.0	77.0	115.3	57.5	57.8	49.9	50.1
1983	124.9	37.2	87.7	29.8	70.2	129.2	64.3	65.0	49.7	50.3
1984	150.2	52.4	97.7	34.9	65.1	154.6	73.9	80.8	47.8	52.2
1985	186.6	70.8	115.9	37.9	62.1	184.5	83.7	100.8	45.3	54.7
1986	226.0	91.7	134.4	40.6	59.4	233.1	96.2	136.9	41.3	58.7
1987	236.9	90.6	146.3	38.2	61.8	244.8	103.2	141.7	42.1	57.9
1988	262.8	104.6	158.2	39.8	60.2	270.7	106.0	164.6	39.2	60.8
1989	294.8	110.5	184.2	37.5	62.5	304.0	110.5	193.5	36.4	63.6
1990	331.3	136.8	194.5	41.3	58.7	345.2	137.3	207.9	39.8	60.2
1991	358.3	137.0	221.3	38.2	61.8	379.4	149.7	229.6	39.5	60.5
1992	415.3	164.9	250.4	39.7	60.3	439.0	181.8	257.2	41.4	58.6
1993	508.8	169.7	339.3	33.4	66.7	528.7	195.7	333.0	37.0	63.0

* Based on Chinese definition.
[†] Revenues collected by the center or localities (before tax sharing), not revenues available to the center and the localities after tax sharing.

SOURCE State Statistical Bureau (1992a, 1994).

1980s, the central government has been increasingly concerned with the potential political and economic consequences of its weakening fiscal power and has repeatedly expressed the intention to increase two ratios – the ratio of government revenue to GNP and the ratio of central government revenue to total government revenue.[8] Although this was written into the Eighth Five Year Plan for 1991–5, the declining trend of the two ratios continued between 1990 and 1993. At the end of 1993, the central government announced a major reform plan which was to change the fiscal contract system into a tax assignment system in 1994.[9] The main objective of this reform was to reverse the trend of the declining two ratios, particularly the ratio of central government revenue to total government revenue, by replacing a negotiation-based system with a rule-based system.

DIVISION OF EXPENDITURE RESPONSIBILITIES BETWEEN THE CENTRAL AND LOCAL GOVERNMENTS

Before 1980, almost all revenues collected by the localities were submitted to the center and all expenditures were financed by central government transfers. Since 1980, as a part of the decentralization in fiscal management, the central government no longer stipulates individual items of expenditure in provincial budgets but continues to formulate the guidelines indicating the acceptable level of expenditure for each province.[10] The allowed expenditure of each province is formulated based on the 'base figure' – the expenditure of the province in a certain base year – adjusted by factors such as changes in policy priority (for example, the State Council's approval of a new development zone may increase this region's expenditure allowance) and inflation (for example, inflation increases the payments of wages and other procurements), price reform (for example, certain price reform requires the provinces to spend more on food and housing subsidies), and so on. The provinces are required to make their expenditure plans consistent with the central government guidelines, but may augment expenditures if more resources are likely to be available to the province. On individual items of local government expenditure, the center has a set of detailed regulations restricting government employees' wages and the items that government agencies (at all levels) can purchase.

The state budget consists of the central budget and local budgets. The local budget includes the provincial, county, and municipal budgets. In principle, the division of expenditure responsibilities between the center and the local governments, in so far as they are limited to the budget, is broadly defined as the following. The central government's expenditure responsibilities include capital construction and technical upgrading of centrally owned state enterprises and outlays on their new product development; national defense; foreign aid and external relations; agriculture, forestry, and water conservation at the central level; industrial, transportation, and commercial operations; education, culture, health and social services; outlays on centrally obligated price subsidies; geological surveys; and repayment of public debt. Local expenditure responsibilities include basic construction and technical improvement of locally owned state enterprises and their new product development; rural production assistance, agricultural development and water conservation at the local level; urban maintenance and construction; education, health, culture, and social services; social welfare and pensions; administrative expenditures; and a range of price subsidies.[11]

There is a substantial overlap between the center's and the localities' responsibilities in actual expenditures. This overlap is common in heavy industrial sectors (for example, electricity and raw materials), as well as in transportation, telecommunications, education, science, and health care. For example, the central government finances electricity plants, railways, and telecommunication networks, and provinces are often required to co-finance these projects. There are both central and local universities, making education both a local and a central responsibility, and a highway or a bridge can be financed by the central government, by the provincial government, or by the municipal government. Since a substantial part of investment outlays come from the central budget, the location of projects becomes a matter of importance. Provinces compete with each other to garner a greater share of these outlays. Both before and after the state budget plan is formally approved by the Congress, intensive lobbying from the local governments for allocating more projects in their own jurisdictions takes place. One typical form of such lobbying is that the local leaders seek help from a senior leader in the central government or the Central Party Committee, and this senior leader writes a note to the State Planning Commission suggesting consideration of a project in favor of this region. According to staff from the MOF, virtually every project approved in the budget plan or added to the plan later was backed by some kind of 'note'. Although there are official guidelines for allocating projects, such as achieving a balance between production and consumption, between industry and agriculture, between heavy and light industry, and between coastal and inland areas, these criteria are usually very vague and the actual allocation mostly reflects the relative bargaining powers of the localities. Chapter 3 will examine in detail how local government lobbying affects China's budget-making.

BASIC PROPOSITIONS FOR ASSESSING CHINA'S FISCAL MANAGEMENT SYSTEM

Economic theories of fiscal management and intergovernmental relations have provided some important guidelines for assessing the structure and performance of the Chinese central–local fiscal relations. We adopt several basic propositions that are most relevant to the issues addressed in this chapter.[12] These propositions are outlined as follows.

Tax policy instruments of the central and subnational governments

The central government should have effective control over adequate fiscal resources in order to perform its role of stabilization, redistribution, and provision of public goods and services. The tax policy instruments that belong to the central government should include rates and bases of the taxes assigned to the center, as well as public spending. The control over these tax policy instruments (tax rates and tax bases) should be realized through setting up the center's own tax collection machines. The subnational governments should be given the right to determine the rates of those taxes which are assigned to them.

Division of tax bases between the central and subnational governments

Musgrave (1983) used equity (consistency of revenue means with expenditure needs) and efficiency (minimizing resource cost) criteria to suggest the following broad principles in tax assignment.

(1) Progressive redistributive taxes should be centralized.
(2) Taxes suitable for economic stabilization should be centralized; lower level taxes should be cyclically stable.
(3) Unequal tax bases between jurisdictions should be centralized.
(4) Taxes on mobile factors of production should be centralized.
(5) Residence-based taxes, such as excises, should be centralized.
(6) Taxes on completely immobile factors should be levied by local authorities.
(7) Benefit taxes and user charges can be levied appropriately by all levels.

Shah (1994) provided two alternative criteria for tax assignment.

(1) The level of government likely to have the best information on a tax base would be the level responsible for levying taxes on such a base.
(2) Revenue means should be matched as closely as possible to revenue needs. Thus, tax instruments to further policy objectives should be assigned to the level of government having the responsibility for such as service.

Expenditure responsibilities of the central and subnational governments

Public goods and services provided by the central government should be restricted to those whose benefits are of national scope or those with significant interregional spill-over effects. Typical examples of these public goods and services include national defense, legal framework and courts, telecommunications, postal services, national roads and highways, and civil aviation. Macroeconomic stabilization and income redistribution are two special 'national public goods'. Therefore, expenditures on items such as unemployment benefits and the social safety net are typically assigned to the central government.

Private goods and services should be provided by the private sector and public goods and services benefitting only a local community should be provided by the local government. However, for many goods and services, such as education, research and development, and environmental protection, it is often difficult to determine the significance of their spill-over effects. In such cases, a careful examination of the costs and benefits of a centralized or decentralized provision of public goods is necessitated in order to balance the objectives of controlling spill-over effects and ensuring efficient project implementation.

While different countries have adopted different approaches to the determination of expenditure responsibilities for the central and subnational governments, one principle for the responsibility division holds universally: when revenue collection is decentralized to the subnational governments, the expenditure responsibilities should be decentralized accordingly. Otherwise, national deficits will be an inevitable result of such a decentralization.

Rules rather than discretion in intergovernmental fiscal relations

The revenue-sharing formula, the tax assignment method, and the division of expenditure responsibilities between the central and subnational governments should be relatively stable in order to avoid adverse incentives to the latter. If a subnational government makes a high tax collection effort and thereby faces a lower revenue retention rate or a higher expenditure responsibility, then its optimal reaction would be to reduce its tax effort. Therefore, *ad hoc* adjustments of the revenue-sharing formula, the tax assignment method, and the division of expenditure responsibilities should be avoided.

Many of the above propositions will serve as general guidelines for this chapter's evaluation of the effectiveness of China's fiscal management system.

THE LACK OF FISCAL POLICY INSTRUMENTS OF THE CENTRAL GOVERNMENT

Many studies on China's central–local fiscal relations point out that the declining ratios of revenue to GNP and the central government revenue to total government revenue were the major problems of the fiscal contract system. They base this argument on a comparison of the Chinese figures with those in other countries.[13] The Chinese government also stated clearly in the Eighth Five Year Plan and the Decision of the 14th Party Plenum that the fiscal reform should proceed in the direction of increasing these two ratios.[14] This section argues that simply looking at the two ratios is not enough for assessing the performance of fiscal management. As countries differ vastly in size, population, and level of development, as well as political and cultural background, the optimal ratios could be very different in different countries. To evaluate the performance of a fiscal management system, in particular the intergovernmental fiscal relations, the key is to look at whether the system enables the central government to use fiscal policy instruments effectively to perform its functions. This will be the task of this and the following sections.

The original purpose of introducing the fiscal contract system was to ensure the center's income while giving the regions more incentive to collect revenues. However, under the fiscal contract system, the localities largely controlled the effective tax rates and bases through controlling tax efforts and the channels through which taxes were collected, although legally the center controls the rates and bases of all taxes. As a result, the center had to resort to various *ad hoc* instruments to influence revenue remittance from the localities and the use of these instruments led to perverse reactions from the localities. In addition, the center's flexibility in using expenditure policy was undermined by the lack of centrally controlled financial resources and the heavy burden of 'capital constructions'. The remaining part of this section will discuss these issues in order.

The first way in which local governments controlled effective tax rates and tax bases was to control their tax efforts, mainly reflected by their use of tax concessions.[15] Nominally, all major taxes were controlled by the central government and tax rates were uniform across regions. Tax rates and tax bases were set by the center and only the center had the authority to abolish old taxes, introduce new taxes, and change tax rates. For central-fixed and shared taxes, only the center could offer tax concessions and exemptions. However, under the fiscal system before 1994, the only binding constraint that a locality actually faced was the fiscal contract with the level of government directly above. As long as a locality fulfilled the revenue target specified in the contract, it tended to use exemptions as much as possible to keep the resource in that locality, even though many of the exemptions were illegal. The rationale was that most of the efforts made for collecting revenues above the target would benefit not just the locality, but also the center. Moreover, no real, implementable, punitive measure were taken against localities' illegal tax concessions.

In 1992–3, a large number of localities began to offer tax concessions that were not authorized by the center. All 30 provinces and thousands of counties and townships launched their own opening up program by establishing various 'zones'. By mid-1993, there were 1800 zones (including special economic zones (SEZs), economic development zones, economic and technology development zones, high- and new-tech zones, special zones for Taiwanese investors, border trade zones, and so on) at and above the county level. Most of these zones were set up by the provincial or county authorities and some even set up by towns and villages. There are no official statistics about 'zones' lower than the country level, but the number was expected to be surprisingly large.[16] Most of these zones offered preferential tax policies that had previously applied in the SEZs, that is, a 15 per cent flat rate on corporate income, a 2 year tax exemption from the first profit-making year, and a 3 year 50 per cent tax reduction afterwards. Some zones offered even more aggressive tax reliefs, a standard form being 5 year tax exemption and 5 year tax reduction by half.[17] While tax relief policies in many of these zones were offered only to foreign investors, there were many other zones, in particular those at the county and township levels, that offered the same policy to domestic investment.

Most of these tax exemptions and reductions were not approved by the central taxation authorities and were considered as illegal by the center. Since mid-1992, the central government has repeatedly claimed that tax relief policies offered by these locally approved zones must be stopped. A circular issued by the center required each province to 'cleanse' these unapproved zones and all promises made by the localities regarding tax concessions could only be fulfilled by using the localities' own budgetary funds. However, it was not until the State Council's September 1993 adoption of a serious austerity program to cool down the economy did these measures take effect.[18]

In addition to 'zones', many regions offered tax concessions under other names. These tax concessions included legal offers made according to state regulations and others which violated state regulations. The central–provincial revenue-sharing system stipulated that provinces could offer tax exemptions and reductions only on local-fixed items, while exemptions or reductions on central-fixed and central–local shared items were subject to approval by the center, except for those areas where special policies applied.[19] However, these special policies were often abused by the local governments.[20] Many localities, if attempting to offer a tax concession, could always find some regulations that were 'applicable'. In addition to using policies promulgated by the center, many local governments offered tax relief simply under the name of implementing local industrial policy or 'subsidizing enterprises that are in difficulties'.

The second way in which local governments controlled effective tax rates and tax bases was to control the channels through which revenues were mobilized. If revenue was collected in the form of a tax, the rate and base were subject to central government control and the proceeds might be shared by the center. If revenue was mobilized through tapping local extra-budgetary funds, it was totally immune from the central government's influence. When regions shifted budgetary funds to extra-budgetary funds, the center's ability to control revenues and expenditures was seriously undermined. In other words, the extra-budgetary funds were fiscal resources circulating 'outside the body' – that is, beyond the control of the central government.

Local extra-budgetary funds include the following two categories. The first is a set of taxes and charges that are controlled by local finance bureaus (not a part of the state budget), the most important of which is the public utility surcharge – a 10 per cent tax on the utility bills of consumers. There are also some minor tax surcharges in this category, including the surcharge on the agricultural tax, revenues received from public housing and public property, and so on. The second category includes the retained earnings by local state-owned enterprises (SOEs) in the form of technical transformation funds, depreciation funds, and special purpose funds and, secondly, various kinds of income of government agencies, including road maintenance fees, agricultural–forestry–water income, income of enterprises owned by government agencies, real estate management fees, and school tuition.[21] Nominally, most of the funds as specified in the second category are controlled by the enterprises, but in recent years local governments have increasingly used various fees and charges to tap state enterprises' retained earnings for government expenditures.

Between 1978 and 1992, the ratio of local extra-budgetary funds to local budgetary funds rose from 66 : 100 to 96 : 100, indicating an increasing/declining importance of extra-budgetary/budgetary funds.[22] In 1993, local governments' tapping local enterprise resources through *ad hoc* charges became so widespread that one enterprise could face more than a hundred fees and charges. The effective tax burden[23] (including the burden of *ad hoc* fees) was as high as 90 per cent in some regions.[24] The central government had to issue several directives requesting local governments to cleanse those 'unwarranted' fees and charges.

As the localities controlled effective tax rates and bases under the fiscal contract system, the central government increasingly relied on a number of non-standard fiscal policy instruments in order to influence revenue remittance by the localities. These instruments include the following.

(1) *Ad hoc* revision of fiscal contract. These revisions were usually at the advantage/disadvantage of provinces with revenue deficits/surplus.

(2) 'Borrowing' revenues from the localities. These borrowed revenues were never returned. Since 1980, this type of borrowing occurred at least twice and the amounts borrowed were added into the regions' remittance requirement.
(3) Arbitrarily transferring expenditure responsibilities to the localities. This was, in effect, equivalent to an increase in localities' revenue remittance from the localities' point of view.
(4) Reclaiming ownership of locally owned enterprises in high-growth industry, so that revenues from these enterprises became part of the central-fixed revenues.
(5) Forcing the localities to 'buy' centrally issued bonds at artificially low prices and repeatedly extending the bonds' term of maturity.[25]

These *ad hoc* instruments had serious negative effects on the regions's tax collection behavior. One of the most important reactions from the regions to such *ad hoc* adjustments was the under-provision of tax efforts. This conclusion is shown rigorously in a game theoretic model in Chapter 2. In this game, the center chooses a revenue-sharing schedule (the percentages of locally collected revenues going to the center), while the regions decide the amount of revenue to collect (the tax effort). As the center is not able to commit to its pre-announced revenue-sharing schedule (that is, it has various *ad hoc* instruments to adjust the revenue-sharing methods before the contracts expire), it tends to appropriate a higher percentage of the locally collected revenue if a region's revenue grows quickly. By doing so, the center can fulfill its objective of equalizing retained revenues across regions. Knowing that the center may renege on its pre-announced policy *ex post*, the regions will react strategically by reducing tax efforts in the first place. This type of strategic reaction explains, at least partially, China's rapidly declining revenue: GNP ratio over the past 15 years.[26]

While revenue collection has been decentralized, the center's expenditure responsibility has not been reduced accordingly. This leaves limited flexibility for the central government to use expenditure policy. The major cause of this problem is that the center is still burdened by 'capital constructions' (that is, investments in productive sectors). These investments are mainly distributed in the energy, raw materials, transportation, and agricultural sectors. In the 1992 budget, the share of capital construction in the central government expenditures still accounted for 32.8 per cent.[27] Many of these projects produce essentially private goods. In market economies, most of these investments are made by the private sector and some by the subnational governments. From the viewpoint of the standard public finance literature, there is no rationale that the center be involved in

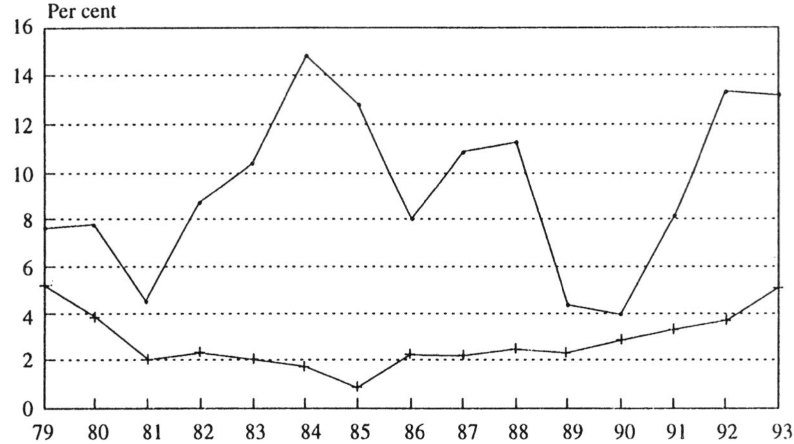

FIGURE 1.1 *Deficit : GNP ratios and real GNP growth rates, 1979–93.*

many of these projects, at least not to this extent. While some of these investments can be justified by cross-regional externalities (such as in the case of roads and railways), most of them are not.

The particular reasons for the Chinese central government to take up these projects include the following.

(1) Some of these projects have to be financed by the central government because the center at the same time imposes price control which makes the sector unprofitable (such as coal mining).
(2) Some industries are extremely profitable, but the entry to such sectors is restricted by the government (such as the electricity sector). Due to administrative entry barriers, the shortage of production capacity is severe and the center is forced to fund projects in these sectors.
(3) The center still controls the major financing channels (it controls credit allocation and restricts bond and stock issuance and the development of private banks), forming another entry barrier to many sectors such as the steel, chemical, and automobile industries in which projects need large initial fixed investment. Given these barriers, the center is again obliged to provide investment to fill the supply–demand gap in the 'bottle-neck' sectors.
(4) The political set-up in China mandates that the central government has to protect employment in SOEs, in particular those subordinate to the central government.

It is clear that most of the reasons that explain the center's engagement in private good production are related to distortions in areas other than fiscal policies, such as planning control, price management, banking regulations, and, most importantly, state ownership of enterprises. Obviously, the reduction of central government expenditure through decentralization can only be achieved with the coordination of reforms in these areas. For example, price liberalization would increase the incentive for enterprises and local governments to invest in many 'bottle-neck' sectors, the removal of the state monopoly in certain sectors would solve the shortage problem and ease the burden of the center, the liberalization of the financial sector might channel more private funds to large projects with high initial capital investment, and privatization and a hardened budget constraint would eliminate the need for the center to subsidize many SOEs.

IMPAIRED PERFORMANCE OF THE CENTRAL GOVERNMENT'S FISCAL POLICY

This section discusses the performance of the central government's fiscal policy with the focus on its link with the central–local fiscal contract system. We will argue that with declining fiscal resources due to the lack of control over tax rates and bases, and with little flexibility in manipulating expenditures, the central government's fiscal policy has been seriously constrained in achieving goals of macroeconomic stabilization and equalization across regions.

Fiscal policy and macroeconomic stabilization

Over the past 15 years, the government's fiscal policy had played virtually no stabilization role. From Table 1.1, one can see that the government revenue : GNP ratio shows a rapidly declining trend with no counter-cyclical pattern, an indication of the fiscal policy's lack of stabilization role. Since the deficit : GNP ratio is often used to measure the degree of fiscal expansion, we can compare the deficit : GNP ratios with real GNP growth rates to detect the correlation between the tightness of fiscal policy with GNP growth. A fiscal policy that stabilizes the economy can be reflected by a negative correlation between the two variables. However, Figure 1.1, which plots the real GNP growth rates and deficit : GNP ratios between 1979 and 1993, shows neither a positive nor negative correlation between the two indices. For confirmation, we conducted a simple linear regression of deficit : GNP ratio on the real

GNP growth rate.[28] The R^2 of this regression is only 0.001 and the *t*-ratio of the slope coefficient is insignificant.[29]

In reality, the management of aggregate demand has been the task of the People's Bank of China (PBC, or the central bank), not the central fiscal authority. While the fiscal authority does not contribute to the economy's stabilization, in contrast, it shifts the burden to the monetary authority in a variety of ways. Many of the unprofitable 'key' projects (such as the projects in the agriculture, infrastructure, and energy sectors) that are supposed to be financed or subsidized by the fiscal budget, are designated to the specialized banks. In addition, the government borrows from the central bank to finance part of its fiscal deficits. The MOF's borrowing from the PBC increased from 27.5 billion yuan in 1985 to 158.2 billion yuan in 1993.[30] The 1993 borrowing was equivalent to 31 per cent of the total government revenue. The borrowing by the fiscal authorities from the central bank exerted strong pressures for the expansion of base money.

The fiscal contract system affected the stabilization role of the center's fiscal policy in two ways. First, the contract system built in the tendency of a declining share of the center's revenue in total government revenue, while at the same time the center did not decentralize the expenditure responsibility to a similar extent. One reason for the center's declining share of revenue explained in the previous section was that the localities reacted strategically to the center's discretionary policy change by lowering their tax efforts and shifting budgetary funds to extra-budgetary funds. Besides this, the revenue-sharing formulae between the center and many provinces implied that the center's revenue growth must be lower than that of the provinces' retained revenues. For example, between 1988 and 1993, ten provinces adopted the type 1 contract, named 'contracted revenue increase' (see p. 7), which allowed the provinces to retain a specific portion of the revenue if the revenue growth rates ranged from 3.5 to 6.5 per cent. Any revenue above the 6.5 per cent growth was fully retained by the provinces. This contract dictated that the share the center received would continue to shrink as long as the provinces' revenue growth exceeded 6.5 per cent. While the share of fiscal resources available to the center fell rapidly, the center retained many of the expenditure responsibilities that were supposed to be decentralized to localities and enterprises, mainly in the form of 'capital construction'. The center's expenditure plan was always overwhelmed by short-term objectives other than stabilization, such as financing 'key' investment projects, paying wage bills for expanding numbers of government employees, and subsidizing loss-making SOEs. The center constantly found itself in a very difficult position of balancing various demands for expenditures, and there was very little room for using expenditure policy to stabilize the economy.

Second, the central–local contract system determined that the center's revenue was unlikely to be strongly pro-cyclical. In the 1988–93 fiscal contract (type 5 contract) with Shanghai, the single largest revenue source of the center's revenue, the transfer from Shanghai to the center was fixed regardless of Shanghai's actual income (see p. 7). The type 1 contract (contracted revenue increase), adopted by ten rich provinces during 1988–93, yielded a less than unity elasticity of the center's revenue relative to the provinces' revenue collection. Although some other provinces shared revenues with the center at a fixed proportion (type 2 contract), these provinces' revenues might not be clearly pro-cyclical due to inadequate tax efforts. As a result, the revenue growth rate did not always move in the same direction as that of the national GNP.[31]

Fiscal policy and redistribution across regions

The central government's role in redistribution has been substantially weakened over the course of decentralization, mainly due to the sharp decline of the center's fiscal resources relative to those of the provinces. As the central government's share of revenue in total government revenue fell (for reasons related to regions' tax efforts under the contract system), both the net transfers from wealthy provinces to the center and the net transfers from the center to poor provinces declined rapidly. Table 1.3 shows 28 provinces' intensity of net transfer to/from the center, defined by (revenue – expenditure)/revenue, in 1983 and 1991.[32] The coefficient of variation (in absolute value) of this intensity dropped from 3.05 in 1983 to 1.89 in 1991. Figures 1.2 and 1.3 plot this index for each of the provinces in 1983 and 1991, respectively, showing a clear trend of increasing fiscal autonomy or a decreasing role in redistribution by the central government. Shanghai, for example, remitted as much as 88 per cent of its revenue to the center in 1983, but only 47 per cent in 1991. Under a system with less intergovernmental transfers, provinces (such as Guangdong) capable of mobilizing large extra-budgetary funds and foreign investments naturally grew faster than others.

For revenue surplus provinces, the center set different revenue-sharing formulae with different provinces to achieve equalization: financially weak provinces were allowed to retain more of the revenues, while the opposite was true for financially strong provinces. However, shared revenues grew at a lower speed than the actual fiscal capacity (for example, as measured by GDP). Consequently, the center found that changing the revenue-sharing methods became less and less effective in addressing the problem of fiscal disparity. For deficit provinces, the central government contracted subsidies (net transfers) to them. The relative size of these transfers has decreased over

TABLE 1.3 *Revenues and expenditures of provinces, 1983 and 1991 (billion yuan)*

	Revenue		*Expenditure*		*(Revenue – Expenditure) Revenue*	
	1983	*1991*	*1983*	*1991*	*1983*	*1991*
Beijing	39.84	89.77	19.61	80.73	0.51	0.10
Tianjing	38.74	58.09	20.49	47.47	0.47	0.18
Hebei	36.39	99.30	28.27	97.30	0.22	0.02
Shanxi	24.15	72.76	24.01	76.89	0.01	–0.06
Inner Mongolia	6.99	39.40	22.83	66.62	–2.27	–0.69
Liaoning	67.88	161.54	34.17	151.45	0.50	0.06
Jilin	14.12	62.46	19.41	79.12	–0.37	–0.27
Heilongjiang	21.56	94.74	30.71	110.07	–0.42	–0.16
Shanghai	153.70	191.88	19.03	101.52	0.88	0.47
Jiangsu	74.61	143.29	32.29	128.18	0.57	0.11
Zhejiang	41.79	121.26	21.94	97.32	0.47	0.20
Anhui	22.39	54.23	20.38	84.97	0.09	–0.57
Fujian	12.37	69.70	17.55	78.13	–0.42	–0.12
Jiangxi	13.53	50.55	17.27	64.68	–0.28	–0.28
Shandong	51.30	142.97	32.41	141.95	0.37	0.01
Henan	36.49	104.73	30.06	108.32	0.18	–0.03
Hubei	40.44	95.09	28.32	99.53	0.30	–0.05
Hunan	29.27	96.13	25.31	101.84	0.14	–0.06
Guangdong	36.33	192.09	37.65	182.48	–0.04	0.05
Guangxi	13.89	61.47	18.84	75.89	–0.36	–0.23
Sichuan	41.38	148.35	36.64	162.86	0.11	–0.10
Guizhou	8.27	45.62	15.55	55.88	–0.88	–0.22
Yunnan	17.17	99.78	24.03	110.82	–0.40	–0.11
Shaanxi	14.54	54.17	18.81	67.31	–0.29	–0.24
Gansu	10.90	39.98	15.53	51.32	–0.42	–0.28
Qinghai	1.54	8.79	7.39	18.23	–3.80	–1.07
Ningxia	1.78	8.30	6.95	17.20	–2.90	--1.07
Xinjiang	5.63	26.47	18.61	52.53	–2.31	–0.98
Mean	31.32	86.89	23.00	89.66	–0.37	–0.19
SD	18.43	48.39	8.34	41.89	1.13	0.36
COV	0.59	0.56	0.36	0.47	–3.05	–1.89

SOURCE MOF (1988, 1992).

NOTE: revenue refers to total revenue collected by a province before tax sharing. It does not include transfers from the center, nor excludes its transfer through tax sharing to the center. Expenditure refers to total expenditure incurred in this province. It equals local revenue collection plus transfer from center and minus its transfer to the center. SD, standard deviation; COV, coefficient of variation.

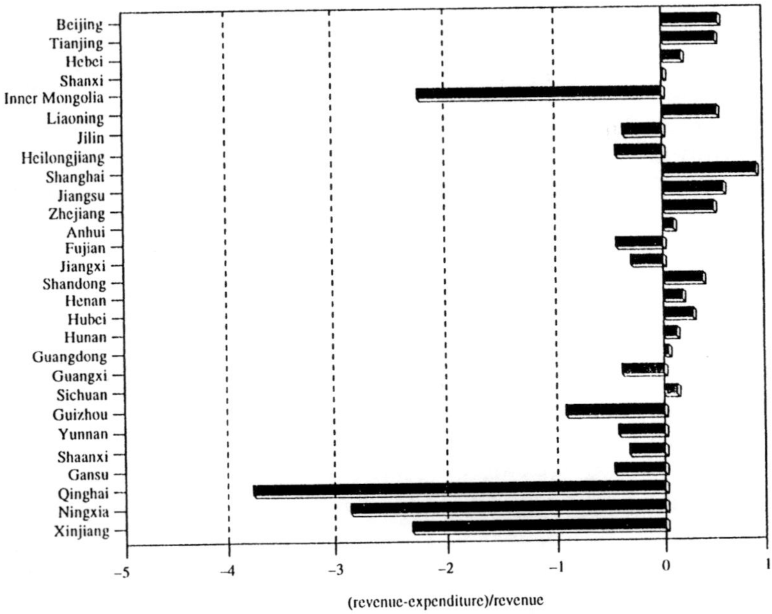

FIGURE 1.2 *Intensity of net transfers from/to the center, 1983.*

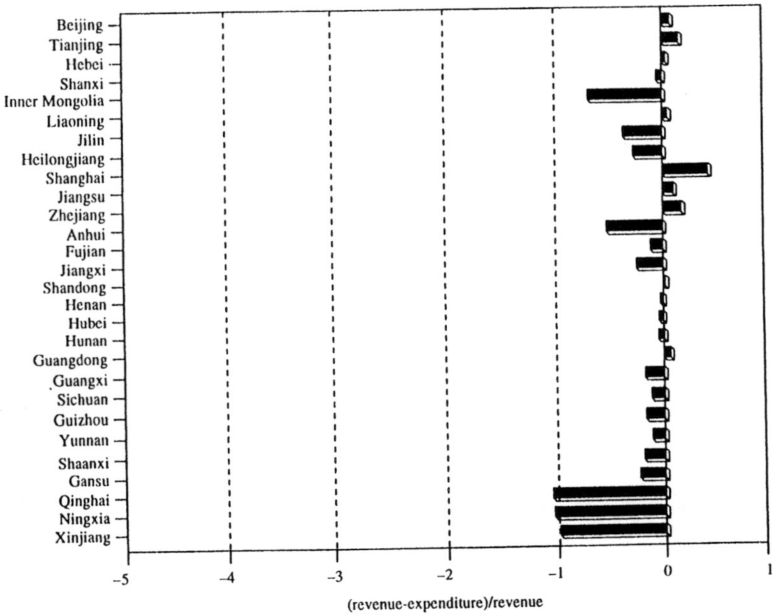

FIGURE 1.3 *Intensity of net transfers from/to the center, 1991.*

the past decade due to the center's increasingly difficult fiscal position. For most poor regions, the center has offered in recent years fixed subsidies in nominal terms, which implied declining subsidies in real terms. In 1985, the center began to increase earmarked grants relative to contracted subsidies. Of these earmarked grants, a large part went to rich regions such as Shanghai and Guangdong, mainly in the form of price subsidies and subsidies to loss-making state enterprises. This shift further reduced the redistribution role of intergovernmental transfers in redistribution.

THE 1994 REFORM: THE TAX-ASSIGNMENT SYSTEM

At the beginning of 1992, the Chinese central government initiated the experiment of a tax-assignment system in nine cities in an attempt to raise the 'two ratios': the ratio of government revenue to GNP and the ratio of central government revenue to total government revenue. In late 1993, as part of the comprehensive fiscal reform, the central government decided to extend this tax-assignment system, with some modifications, to all 30 provinces, starting in 1994. The main elements of this reform plan were the following.[33]

Expenditure

The central government would be responsible for national defense, diplomatic affairs, the armed police forces, key construction projects, repayment of principal and interest on domestic and foreign debts of the central government, and the spending of the central government administrative departments, while other expenditures would be undertaken by local governments.

Tax assignment

As before, taxes would be divided into central-fixed taxes, local-fixed taxes, and shared taxes. The central-fixed taxes would mainly include customs duties; consumption tax collected by customs; consumption tax on items such as cigarettes, liquor, and gasoline; value added tax (VAT) and income tax of centrally owned state enterprises; turnover taxes of railway, banks, and insurance companies; and income tax from financial institutions set up by the headquarters of the PBC. Taxes collected by local governments (that is, the local-fixed taxes) would mainly include business tax (except for the turnover taxes of banks, railways, and insurance companies), income tax of local enterprises, and personal income tax. The shared taxes would mainly include

VAT, securities trading tax, and natural resources tax. VAT would be divided at 75 per cent for the central government and 25 per cent for the local governments. The securities trading tax, currently collected only in Shanghai and Shenzhen in the form of stamp duty, would be divided at 50 : 50 between the central and local coffers. Natural resources tax would be mainly kept by the local governments for the time being.

One major difference between the new system and the old system is that, instead of allowing the localities to collect virtually all the taxes, the central government would set up its own tax collection agency, namely the National Tax Service, responsible for collecting both the central-fixed taxes and shared taxes. The Local Tax Services would collect only the local-fixed taxes.

It was stated as an objective of this reform that the center would gradually control over 60 per cent of the total revenue, of which approximately about 40 per cent would be the expenditure of central government. The other 20 per cent, out of the 60 per cent central government revenue, would be allocated to local governments through central government grants.

Tax bases and tax rates

The bases and rates of various taxes would be adjusted substantially in the new system. The tax reform would cut income tax rates for large- and medium-sized SOEs from 55 per cent to a uniform 33 per cent. This rate would be applied to all types of enterprises regardless of ownership.[34] The number of commodities that would be subject to VAT would be substantially increased, and for most products the VAT rate would be 17 per cent, a rate higher than those applied before. A business tax of 3–5 per cent would be applied to firms in the service sector, such as entertainment, food, insurance, financial, and transport firms. A consumption tax would be levied on a dozen items, including cigarettes, liquor, and gasoline. Tax on personal income would be progressive, beginning at 5 per cent on salaries of 800 yuan a month and rising to a maximum of 45 per cent. Foreigners and the Chinese would be taxed at the same rates. The new tax system would also introduce five additional categories of taxable income, including income from property transfer, freelance writing, gains from real estate and stock transactions, inheritance, and donations.

This reform would be carried out in a progressive way in 1994 and afterwards. The vested interests of the local governments under the old system would be guaranteed. That is, the revenue-sharing formula under the new system was designed in such a way that a province would retain no less revenue than it did in 1993. The retained revenue of a province in 1993 would be used as the basis for calculating the amount of shared revenues returned from the central government to the province since 1994.[35] The center would increase its share in total government revenue only from the increase in the shared revenues, mainly from VAT. To win the support of the local governments, the center also promised to allow, during 1994–5, the continuation of tax exemptions approved by provincial level governments, and not to shift new expenditure responsibilities to the localities.

The adverse reaction of the localities to this transitional arrangement (which was announced in November 1993) was to collect as much revenue as possible in order to increase the base for retained revenues over the coming years. Some localities urged the local banks to make loans to enterprises to pay previously unpaid taxes. Some localities even collected taxes in advance (that is, collected part of 1994's tax). The last quarter of 1993 witnessed an increase in the total revenue collection by approximately 40 per cent compared to the same period of 1992.[36] The center believed that this type of excessive taxation for the purpose of bidding up the quota for future retention would increase the burden of the enterprises and reduce their growth potential. The center threatened that if the localities collected excessive taxes or collected taxes in advance, a circular would be issued to stop such practices, the regions' base figure would be re-evaluated, the income would be confiscated, and those officials involved would have to take the blame.

IMPLICATIONS OF THE 1994 FISCAL REFORM AND REMAINING PROBLEMS TO BE ADDRESSED

It is still early to assess the overall consequences of the 1994 fiscal reform. However, a number of implications regarding the central government's ability to use fiscal policy are clear based on our analysis in the previous sections. This section briefly discusses the following questions: to what extent has the new reform addressed the problems arising from the center's lack of fiscal policy instruments? What are the remaining problems to be addressed?

Firstly, the new system will, to a large extent, eliminate the localities' *de facto* control over effective tax rates and bases, and provide the center with tax policy instruments immune from the localities' strategic reaction. Under the old contract system, the localities performed most of the tax collection function and then shared revenue 'upward' with the central government. This system gave the localities an important instrument to play a game with the center: the tax collection effort. The localities therefore had substantial bargaining power in the central–local negotiations since they could react to the center's policy strategically by changing tax efforts and shifting budgetary funds to extra-budgetary funds. It was mainly for this reason that the outcome of the contract system had been much less predictable than the center originally expected. In the new system, the National Tax Service will collect the shared revenue and then share the proceeds 'downwards' with the localities. Since the center collects and controls the shared revenues, the local governments' strategic reaction of reducing efforts will no longer exist. In other words, by centralizing the revenue collection of the shared revenues, the new system will eliminate an important source of the localities' incentive problem.

The 1994 reform, if sucessfully implemented, would be an important step in moving the fiscal system towards a rule-based system from the current discretionary system, for there would be no more one-to-one contract negotiation. In the new system, the ratio of revenue sharing is unified across provinces. The fiscal gap is addressed mainly by central government grants, which are, at least nominally, under the control of the center. The new revenue-sharing formula of 75 per cent (center) versus 25 per cent (localities) will substantially increase the center's share in the total government revenue. With the effective control over revenue collection and an increasing share of resources available for intergovernmental transfers, the center will gain more flexibility in using tax and expenditure policies to achieve goals of stabilization and regional equality.

Secondly, future reform should proceed in the direction of establishing rules for intergovernmental fiscal relations through national legislation rather than administrative intervention. So far, the division of tax powers and expenditure responsibilities has been largely determined by the central government, rather than through legislation of the National Congress. This allows the central government to change the rules frequently in its own interest. Without a commitment to a stable division of powers, the regions react strategically to expected policy changes, thereby leading to a very unpredictable outcome. In contrast to the case of China, local autonomy in many federalist countries and even unitary countries such as Japan has been guaranteed by the Constitution. The post-war Constitution of Japan has a chapter on local self-government. This chapter states that regulations concerning the organization and operations of local public entities should be fixed by law in accordance with the principle of local autonomy.

Normally, the constitution cannot specify the details of tax and expenditure assignment to each level of government. This task is left for laws on local taxation, such as the Local Finance Law and Local Tax Law in Japan. Among many laws regarding fiscal issues, Japan's Local Finance Law and Local Tax Law have been the main pieces that govern the division of taxing power and expenditure responsibilities between the central and local governments. The rules specified in these laws have ensured the stability and predictability of local budgetary revenue and expenditure. They have also protected local governments from arbitrary burden-sharing requirements that might otherwise have been imposed by the central bureaucracy. In addition, these rules have helped to develop local governments' trust in the national government. Once the tax and expenditure assignments are written into law, they cannot be easily altered at the central government's discretion. This provides the central government with a device of commitment to stable intergovernmental relations. With such a commitment, the local governments' behavior in taxation and expenditure becomes more predictable as strategic reactions to possible changes in the central–local fiscal arrangement are eliminated. A predictable pattern of local fiscal behavior will, in turn, contribute to the central government's capacity of macroeconomic management.

Thirdly, institutional protection of local autonomy should be established. A legal definition of the central–local fiscal relations is not sufficient to guarantee its strict implementation. Institutions representing the interests of the local governments in the national political arena must be established to counteract the strong tendency of central government agencies to appropriate fiscal powers from the localities.

In Japan, this task is performed successfully by the Ministry of Autonomy. Individual local governments, although numerous, normally do not have the incentive and the capacity to seek the protection of local autonomy. The Ministry of Autonomy stands for local governments as their sole advocate in intergovernmental fiscal relations. Its function is important because local governments have very little formal power over legislative processes in the national Diet (the National Assembly), while central government agencies generally seek more control over local governments. Under these circumstances, the Ministry of Autonomy works as an institutional protector of local governments by requiring central government ministries to obey the stipulated intergovernmental relations in the laws. In particular, it requires that the ministries do not arbitrarily transfer expenditure responsibilities or reduce transfers to the local governments.[37]

Fourthly, division of tax bases according to ownership is not desirable. For the purpose of minimizing revenue redistribution during the transition, the central government decided to retain the current division of fixed revenues from enterprise income taxes by ownership. As in the old system, the central government receives the income taxes from the centrally owned enterprises and the localities receive income taxes from the locally owned enterprises. This division has two major flaws.

(1) The income of the locally owned enterprises is a large and dynamic tax base. Without a claim on the local enterprises' income tax, the center

loses an important instrument in affecting enterprise behavior nationwide. For example, a progressive income tax on all enterprises may provide the center with an automatic stabilizer and income tax credit on these enterprises during recession may be used as a stimulus to the economy.

(2) The center will be tempted to reclaim the ownership of some local firms whenever its fiscal resources seem to be inadequate. This temptation will trigger the regions' incentive problem similar to that caused by contract renegotiation (see p. 17). Without the center's commitment not to reclaim the ownership of any local enterprises, regions with well-managed and profitable local enterprises face a high probability of being victims. Recognizing this, regions may intentionally under-invest in these enterprises or under-report the profits in order to reduce the chance that their local enterprises' ownership will be confiscated.

Fifthly, the division of expenditure responsibilities has yet to be clearly defined. According to the reform plan, the expenditure responsibilities are only divided into two broad categories, the center's and the localities'. However, in reality, there is a substantial overlap of expenditure responsibilities between the center and the localities. The center has exercised discretion in assigning responsibilities to regions when it faces fiscal difficulties. Regions have taken advantage of the center's involvement in the productive sectors by intensely lobbying for grants. The lack of a clear definition of expenditure responsibilities has created incentives for both the center and the localities to direct more resources to bargaining. In addition, if a region that has faster growing local revenues than others is assigned more expenditure burden by the center, this could become a disincentive for the region's tax collection.

Sixthly, the new system will be more transparent if the localities are allowed to determine the bases and/or rates of local taxes. The tax assignment system designates a number of local-fixed taxes to the localities, but the tax bases and tax rates are still determined by the center. From the viewpoint of the fiscal federalism literature, this is not an efficient decentralized system in which the localities can determine their optimal level of tax burden and level of public services in accordance with local conditions. Nevertheless, the localities in China may choose to vary their tax efforts to adjust the effective tax rates and tax bases. Since the enforcement of unified effective local tax rates and bases is difficult anyway, as manifested by the past experience under the contract system, the center's nominal control over local tax rates and bases only weakens the system's transparency, predictability, and stability. We suggest that the center surrenders the tax power for local-fixed taxes – that is, the right to determine local tax rates and bases – to the localities. The center, as long as it controls an appropriate share of total fiscal resources and the instruments of intergovernmental transfers, has no reason to worry about loss of control over local tax rates and tax bases. In contrast, the autonomy of localities in determining their local tax policies may enhance local welfare through establishing a tax structure well-suited to the local residents' preferences, industrial structure and technologies, and market conditions.

Between the federal system in which local governments have the right to determine local tax rates and tax bases and the system adopted by many unitary countries in which local governments act only as tax collectors of the centrally determined local taxes, the Japanese system provides a somewhat intermediate case that reconciles the needs for central government control and local autonomy. Under the Japanese system, the central government sets fixed tax rates for a number of local taxes and provides ranges for some other local taxes. The local governments and local assemblies are given the authority to propose new taxes, and most of these proposals are approved by the center. The major advantages of such an arrangement are as follows.

(1) By keeping uniform rates of certain local taxes and maintaining the right to approve new taxes, the center can avoid or mitigate the problems of excessive local taxation by some local 'leviathan' governments or inadequate taxation due to regional tax competition.
(2) The localities are given some flexibility in choosing tax rates and bases according to their local conditions, including income levels, preference of the local residents, and resource endowments.

Finally, how to design intergovernmental transfer is still an unaddressed question. According to the reform plan, the center will transfer approximately 20 per cent of its revenues to the localities through central government grants. According to the MOF, the Australian and the Canadian grant systems are being studied as potential models. However, no detailed plan of such transfers has been announced. We believe that this is one of the most interesting areas that deserve more attention and careful research. A number of questions require serious answers.

(1) What are the objectives of the intergovernmental transfer? If one of the objectives of the transfer program is to address the problem of underprovision of local public services due to externality, what are these externality problems?
(2) If the objectives include equalization and reduction of vertical fiscal gap, how should the center measure fiscal capacity and expenditure needs?

(3) What types of intergovernmental transfer will be used to address each of the stated objectives?

The experiences of most federalist countries show that a uniform and transparent formula for allocating central government grants to the localities is important in reducing wasteful lobbying activities and in maintaining the system's stability. In addition, with a well-designed formula, an increase in local tax revenue or a reduction in local spending will reduce transfers from the center by less than the same amount, thus preserving the localities' incentive to collect local taxes and control local expenditures.

CONCLUSIONS

This chapter looks at the impact of China's fiscal decentralization on the central government's ability to use fiscal polices in achieving macroeconomic objectives. It is argued that under the fiscal contract system introduced in the early 1980s, the localities controlled the effective tax rates and tax bases in the following two ways. First, they controlled the tax collection efforts by offering varying degrees of tax concessions. Second, they found ways to shift budgetary funds to extra-budgetary funds, thus avoiding tax sharing with the center. As a result, the center had to resort to various *ad hoc* instruments to influence revenue remittance from the localities and these instruments led to perverse reactions from the localities. On the expenditure side, the center failed to achieve corresponding reductions in expenditure when revenue collection was decentralized. The center's flexibility in using expenditure policy was seriously undermined by the lack of centrally controlled fiscal resources and the heavy burden of 'capital constructions'.

Due to the lack of fiscal resources and policy instruments, the central government has found itself in an increasingly difficult position to achieve the goals of macroeconomic stabilization and regional equalization. The adoption of the tax-assignment system in the 1994 fiscal reform was an important step taken by the central government to address these difficulties. Under the new system, the center is recentralizing the administration over collection of shared taxes and will eventually obtain a larger share of fiscal resources due to the new revenue-sharing formula. As a result, the center will significantly improve its ability to use tax and expenditure policies in macroeconomic management. Nevertheless, the new tax-assignment system has yet to address the following problems.

(1) Future reform should proceed in the direction of establishing rules for intergovernmental fiscal relations through national legislation rather than administrative intervention.
(2) Institutional protection of local autonomy should be established.
(3) Division of tax bases according to ownership is not desirable.
(4) The division of expenditure responsibilities has yet to be clearly defined.
(5) The new system still lacks transparency and predictability because the localities are not allowed to determine the bases and/or rates of local taxes.
(6) How to design intergovernmental transfer is still an unaddressed question.

APPENDIX 1.1 MACROECONOMIC MANAGEMENT AND LOCAL AUTONOMY IN JAPAN[38]

The Japanese central government has a number of instruments to control local public finance in order to achieve its macroeconomic goals. The first of such instruments is to delegate functions to local governments. When a local government undertakes an agency-delegated function, it becomes an agency of the central government and executes functions as directed. Agency-delegated functions are essentially the responsibility of the central government but are delegated to local governments for administrative convenience. At present, more than half of the functions performed by the prefectural governments are agency-delegated.

The second important instrument for the central government to exercise macro fiscal control is to set guidelines, standards, and regulations for almost all local activities, even for functions that are inherently the responsibility of local governments. To implement these guidelines and regulations, the conditions attached to various central government disbursements for specific purposes are most instrumental. These disbursements are distributed on the condition that the recipient follow the directives issued by the national governments. If a local government fails to observe national directives, it may be requested to refund the disbursement in whole or in part.

The basic instrument for the central government to achieve regional equality and to standardize the level of local public services is the use of local allocation tax. This program equalizes the fiscal capacity between local governments by supplementing the shortage of tax revenue. This tax enables local governments to provide public services at the level prescribed by the central government. When a local government does not maintain the level prescribed for public services, or has paid an excessive amount for the services, the national government may reduce the local allocation tax for that local government.

Although Japan is a unitary country, local autonomy is guaranteed by the Constitution, in chapter VIII, on local self-government. The Local Autonomy Law, the basic statute concerning the local government system, contains provisions for local tax and finance administration. Various other laws, such as the Local Finance Law, the Local Tax Law and the Local Allocation Tax Law, delineate the finer points of local finance and taxation. The fiscal relationship between the central and local governments is specified in the Local Finance Law: 'The central government shall endeavor to promote the self-dependence and soundness of local finance, and refrain from any action prejudicial to the fiscal autonomy of local governments or from shifting its burden upon local governments' (Ishihara, 1993, p. 143). Under this article, fiscal intervention by a higher level of government is limited to maintain the sound operation of public finance.

A salient feature of the Japanese system in protecting local autonomy is the function of the Ministry of Autonomy. This ministry stands for local governments as their sole advocate in intergovernmental fiscal relations. Its function is important because local governments have very little formal power over legislative processes in the national Diet (the National Assembly) while central government agencies generally seek more control over local governments. Under these circumstances, the Ministry of Autonomy works as an institutional protector of local governments by requiring central government ministries to obey the stipulated intergovernmental relations in the laws. In particular, it requires that the ministries do not arbitrarily transfer expenditure responsibilities or reduce transfers to the local governments.

A clear objective of the Ministry of Autonomy is to obtain as many revenue sources as possible for the local governments to balance their budgets. Each year the ministry formulates a local public finance program that covers all local governments' standard activities except local public enterprise special accounts. On the revenue side of the program, all the standard sources of local revenue – such as local taxes, the local allocation tax, central government specific purpose disbursements, local loans and fees, and so on – are considered. If the estimated program is expected not to balance for the year, the Ministry of Autonomy proposes measures (such as local tax amendments, an increase in the local allocation tax, an increase of local loans, curtailment of local administrative costs, and so forth) to restore the fiscal balance for the local governments.

There are some doubts about how much the Ministry of Autonomy can do: since the Minister of Autonomy is appointed by the Prime Minister, can the Minister of Autonomy resist the pressure from the Prime Minister who may be favoring a proposal of centralization? This concern is certainly legitimate, but there exist mechanisms to prevent this from happening. The local governments, whose interests are represented by the Ministry of Autonomy, have the incentive to protect the Autonomy Minister once he or she is threatened by the top government leaders. The local leaders can express their concerns through direct communications with the Prime Minister, or through the local representatives in the National Assembly. In other words, the Ministry of Autonomy is a representative of the localities when it negotiates with other ministries; at the same time it is backed by the local governments when facing threat from the Prime Minister. A stable equilibrium of power division can therefore be achieved.

2 Central Government Credibility and Local Tax Efforts: a Game Theoretic Model

INTRODUCTION

As discussed in Chapter 1, China introduced a decentralized revenue-sharing system (that is, the fiscal contract system) in 1980. Despite many changes, the fiscal contract system maintained two features between 1980 and 1993: the provincial and lower level governments collected most of the revenues, except for a few taxes such as custom duties; each province remitted a certain amount (or percentage) of its locally collected revenue to the center, or received a certain amount of central governmental transfer, according to a contract signed with the center.

These contracts, despite the variety of forms, in effect determined the share of locally collected revenues going to the center and the share going to the localities.[1] As long as a province fulfilled the fiscal contract, it had the authority to determine how to spend the retained revenues. By leaving a proportion of the locally collected revenues to the provinces, the new system was expected to increase the localities' incentive to collect more revenues.

In contrast to what was expected, however, regions' tax collection efforts actually declined rapidly. Between 1978 and 1993, the ratio of total government revenue to GNP dropped from 35 to 13 per cent (using the standard definition, see Table 1.1). While GNP grew at an annual average rate of nearly 9 per cent during this period, government revenues grew at an average rate of only 4 per cent.[2] The consequences of this change include increasing government deficits, inadequate investment in the government's priority projects, and the inability of the government to carry out reforms that involve significant income redistribution effects. Moreover, difficulties on the fiscal front have forced the government to rely on the banking system for policy lending, which in turn created serious incentive problems that resulted in rapid credit expansion and high inflation.

This chapter argues that the decline in local tax efforts can be explained, at least partially, by the center's lack of commitment to the fiscal contracts.

Nominally, these contracts were good for 3–5 years. In practice, however, the central–local revenue-sharing contracts were repeatedly revised by the central government before the term of the contracts expired. These revisions took many forms, examples of which follow.

(1) Directly renegotiating revenue-sharing contracts.
(2) 'Borrowing' revenues from the localities. These borrowed revenues were never returned.
(3) Arbitrarily transferring expenditure responsibilities to the localities.
(4) Reclaiming ownership of enterprises in high-growth industry, so that revenues from these enterprises become part of central-fixed revenues.
(5) Forcing the localities to 'buy' centrally issued bonds at artificially low prices and repeatedly extending the term of the bond maturity.

A game theoretic model is constructed to examine the interaction of the central government's revenue-sharing arrangement and the representative regions' revenue collection. In this game, the center chooses a revenue-sharing schedule (the percentages of locally collected revenues going to the center), while the regions decide the amount of revenue to collect. It is shown that if the center is not able to commit to its pre-announced revenue-sharing schedule, it tends to revise fiscal contracts to appropriate a higher percentage of the locally collected revenue if a region's retained revenue (in per capita term) is higher than those in other regions. Knowing that the center may renege on its pre-announced policy *ex post*, the regions will react strategically by reducing tax efforts. In practice, the reduction in tax efforts by the Chinese local governments took such forms as abusing tax exemption powers and diverting budgetary funds to extra-budgetary funds.[3]

The picture is different if the center can commit to a pre-announced revenue-sharing schedule. Given the center's commitment, the regions will not react strategically and the revenue collection will be higher than in the no commitment case. The center and the regions will both be better off under the commitment regime. But why does the center not want to commit in reality, as manifested by China's experience? It is demonstrated that the nature of the problem is that the center's pre-announced optimal policy (revenue-sharing schedule) is time inconsistent. Before the regions have collected the revenues, the center finds it optimal to announce a policy that is more favorable (in the sense of a lower revenue remittance requirement) to regions that are more productive (in the sense of lower revenue collection costs) than others, in order to increase total revenue collection. In most cases, the center's concern about efficiency (high revenue collection) leads to an unequal distribution of retained revenues across regions. After the regions have collected their revenues, however, increasing revenue collection is no longer an objective of the center, and the center's only concern is to achieve equalization of retained revenues across regions. Therefore, the center almost always finds it optimal to revise the revenue-sharing schedule *ex post*. In contrast, the suboptimal policy under the no commitment regime is time consistent and better describes the actual situation in the Chinese economy.

The finding of the model suggests that applying a simple rule may solve the time-inconsistency problem. For example, the Congress can legislate that any major revision of revenue-sharing methods must go through certain complicated legal procedures. This will substantially increase the cost of contract revision and, hence, reduce the center's incentive to do so. At the same time, the regions' legal rights for any fiscal contracts with the center should be protected by the law. In addition, the regions should be able to resort to certain institutions when their rights are violated by the center's discretionary policy change.

Besides the proposal of legally restricting the center's ability to revise revenue-sharing methods, there are a number of alternative mechanisms that can help address the center's time-inconsistency problem. First, a center that cares sufficiently about its reputation may have less incentive to renege on pre-announced policies. Second, if the egalitarian center can appoint a less egalitarian finance minister (who gives a higher weight to the productive region and a lower weight to the less productive region) to design and implement the central–local revenue-sharing policy, the outcome will be better even without changing the no commitment system. Third, if the center can base its revenue redistribution on some more general measures of regions' welfare, the outcome can also be improved under the no commitment system.

Although many studies on China's central–local fiscal relations have pointed out the trend of a declining revenue : GNP ratio, most of them remain at the descriptive stage.[4] Very few of them paid attention to the negative impact the central government's unwillingness to commit credibly to fiscal contracts might have on regions' revenue collection. Some authors have pointed out that such a hypothesis could be potentially important, but there was no follow-up study. This chapter is the first attempt to analyze formally the strategic interaction between the center and the localities under the fiscal contract system and its impact on local tax efforts.

The rest of this chapter is organized as follows. The next section describes the model and derives the equilibrium in a hypothetical set-up where regions are cooperative (that is, they maximize the sum of their utilities in playing the game with the center). This set-up yields the highest level of total revenue compared to the case of non-cooperative regions discussed in the second section. The second section describes the more realistic case where the regions are non-cooperative, that is, each disregards the benefits of their revenue collection on other regions. We compare the equilibrium outcomes under two different regimes: one is a no commitment regime under which the center determines the revenue-sharing schedule after the regions have collected the revenues; another is a commitment regime under which the center pre-announces and commits to a revenue-sharing schedule. It is shown that the no commitment regime yields a lower level of revenue collection than a commitment regime under the revenue-sharing system. The third section shows that the crux of the problem is that the center's optimal policies are not time-consistent (that is, the center has the *ex post* incentive to renege on pre-announced revenue-sharing methods after the regions have collected revenues). The fourth section offers a number of ways to address the time-inconsistency problem. The four topics in this section are legally restricting the center's ability to change policies, the center's reputation effect, appointing a 'less egalitarian' finance minister, and a 'benevolent' government. The fifth section looks at the proposal adopted by the Eighth Five Year Plan that recommends that the center increases its share in total government revenue. With the model developed in the fifth section, it is shown that such a proposal will not correct the regions' under-provision of tax efforts and that under certain conditions it may worsen the situation. Concluding remarks are given in the last section.

1. COOPERATIVE REGIONS: THE BENCH-MARK CASE

This section discusses the hypothetical case of the central–local fiscal relations where the regions behave cooperatively, in the sense that each region takes into account the externality of its revenue collection on other regions. This analysis will provide a bench-mark for a comparison with the more realistic case where the regions are non-cooperative, in the sense that each region ignores the externality of its revenue collection on other regions. It will be shown that this bench-mark case yields the highest revenue collection compared to the case of non-cooperative regions discussed in the following section.

We assume there is *one* central government and *two* regions in the economy. Each region has a population of one. The center chooses the tax rates (the percentages of locally collected revenues to be submitted to the center) and the regions make the decision of how much revenue to collect in their jurisdictions.

Let Y_i be the ith region's revenue, where $i = 1, 2$ and x_i be the share of revenue remitted to the center from ith region (it is referred to as the tax rate hereafter).[5] The center and the two regions play a two-stage game. The timing of this game is as follows.

(1) In stage 1, region 1 and region 2 choose Y_1 and Y_2.
(2) In stage 2, after observing Y_1 and Y_2, the center chooses x_1 and x_2.

In the terms of game theory, this is a three-player, two-stage game. In terms that are familiar to many readers, the above assumption states that the center plays a Stackelberg game with the regions and the regions move first. In the game, the regions are the leaders and the center is the follower. This assumption embodies the fundamental feature of the current Chinese fiscal system: the center does not pre-commit to a constant revenue-sharing method and the regions take this absence of commitment into account when choosing their revenues. Between the two regions, we assume in this section that they cooperate with each other and maximize the sum of their utility. This is only a bench-mark case which will be used for purpose of comparison with the more realistic cases discussed in the following section.

The equilibrium is defined by two conditions: the center responds optimally given each region's collected revenues and the regions optimize given the center's reaction function to the regions' choices.

To solve for the equilibrium, we follow the standard procedure of backward induction. That is, we first solve for the center's optimal choices given Y_1 and Y_2 (the center's reaction functions to regions' choices) and then solve for the regions' optimal Y_1 and Y_2 given the center's reaction functions.

First consider the central government's problem. We assume that the central government has the objective of maximizing the sum of the logs of the two regions' after-tax revenues. The log utility function captures the egalitarian spirit of the central government – it dislikes the fact that one region retains more after-tax revenue than the other. The center's tendency to equalize after-tax revenues has been pointed out in many studies (Agarwala, 1992; Bell *et al.*, 1993).

Agarwala (1992) arranged the 30 Chinese provinces (including autonomous regions and municipalities directly under the State Council) in ascending order of per capita revenue collection and grouped them into three classes – 'low', 'medium', and 'high' per capita revenues (see Table 2.1). In the low group, per capita expenditures (after revenue sharing) are higher than per capita revenue collections (before revenue sharing), whereas in the high group, the reverse is true. The extent of equalization is brought about by the comparison of group averages. The gaps between group average per capita

TABLE 2.1 *Interprovincial disparities in per capita government revenues and expenditures, 1985*

Group average	*Per capita revenues (yuan)*	*Per capita expenditures (yuan)*
Bottom one-third ('low')*	55	113
Middle one-third ('medium')	73	117
Top one-third ('high')†	112	108

* Leaving out the extreme observations for Tibet.
† Leaving out the municipalities of Shanghai, Beijing, and Tianjin.
SOURCE Agarwala (1992, p. 18).

expenditures are much narrower than those between average per capita revenues. If one leaves out the extreme observations for Tibet and the municipalities of Beijing, Shanghai, and Tianjin because of their special expenditure needs, the group averages of per capita expenditures are remarkably close to each other.

According to Shah (1991a), the two main economic justifications of intergovernmental fiscal arrangements are efficiency and equity. The efficiency criterion requires intergovernmental transfers (for example, various types of matching grants) to address interjurisdictional spill-over effects (Oates, 1972; Gordon, 1983; Hoyt, 1992). The equity objective requires intergovernmental revenue sharing or transfers to eliminate or reduce differential net fiscal benefits (for example, the levels of public goods or services) across regions. Tax-sharing in Germany, for example, is used for equalization purposes. An important part of the proceeds of VAT and other taxes collected by the states are allocated by the center in such a way as to bring the per capita revenues of all states at least up to 95 per cent of the national average (Agarwala, 1992). In Canada, where fiscal equalization is achieved by central government transfers or grants to the regions, the system of 'equalization grants' is intended to 'augment the fiscal capacity of member provinces up to the national average capacity' (Ma, 1996c). At present, China is studying the Canadian intergovernmental transfer system as a possible model for standardizing its revenue-sharing relations with the provinces.

In this chapter we focus on the equity objective of the central–local fiscal arrangement. The center's objective function is written as

$$U_c = \log(1 - x_1)Y_1 + \log(1 - x_2)Y_2 \tag{2.1}$$

where Y_i is the per capita revenue. The center's revenue should cover an expenditure target E:

$$E = x_1 Y_1 + x_2 Y_2. \tag{2.2}$$

This expenditure target, E, is exogenously given.[6] We assume E is small enough so that $x_1 \leq 1$ and $x_2 \leq 1$.

The center maximizes equation (2.1) subject to (2.2). Solving the first-order conditions gives the following equation:

$$(1 - x_1)Y_1 = (1 - x_2)Y_2 \tag{2.3}$$

that is, the center equalizes the after-tax revenues between the two regions. Combining equations (2.2) and (2.3) gives the optimal levels of x_1 and x_2 as functions of Y_1, Y_2, and E:

$$x_1 = 1/2 - (Y_2 - E)/2Y_1 \tag{2.4}$$

and

$$x_2 = 1/2 - (Y_1 - E)/2Y_2. \tag{2.5}$$

The second-order conditions are satisfied globally. From equations (2.4) and (2.5), the obvious properties of x_1 and x_2 are

$$\partial x_1/\partial Y_2 < 0 \tag{2.6}$$

and

$$\partial x_2/\partial Y_1 < 0. \tag{2.7}$$

These results state that there is a positive externality of each region's revenue collection on the other region: an increase in one region's revenue reduces the other region's tax rate.

Having solved for the center's reaction functions, equations (2.4) and (2.5), we turn to look at the regions' problem. Since in the game the regions act as leaders, they incorporate the tax rates as functions of their own choices, that is, equations (2.4) and (2.5), when solving their problems. Region 1's individual objective function is

$$U_1 = (1 - x_1(Y_1, Y_2))Y_1 - C_1(Y_1)$$

where $C_1(Y_1)$ is the cost incurred in region 1 when it collects revenue Y_1. This cost can be interpreted as the administrative cost of revenue collection. We assume that the cost function has a quadratic form: $C_1(Y_1) = a_1 Y_1^2$. It reflects increasing marginal difficulty in collecting revenues. One will see the critical role it plays in determining an optimal solution of the region's problem; without a cost function the objective function is not a quasi-concave one and thus has no maximum.

The two regions may have different cost coefficients a_1 and a_2. The higher this coefficient, the more costly it is for a region to collect additional revenues. With this cost function, region 1's objective can be written as:

$$U_1 = (1 - x_1(Y_1, Y_2))Y_1 - a_1 Y_1^2, \tag{2.8}$$

and similarly for region 2.

When the two regions behave cooperatively, they maximize the sum of the their individual utilities:

$$U_1 + U_2 = (1 - x_1(Y_1, Y_2))Y_1 - a_1 Y_1^2 + (1 - x_2(Y_1, Y_2))Y_2 - a_2 Y_2^2. \tag{2.9}$$

The choice variables of the two regions are Y_1 and Y_2. Substituting equations (2.4) and (2.5) into equation (2.9), equation (2.9) becomes

$$U_1 + U_2 = Y_1 + Y_2 - E - a_1 Y_1^2 - a_2 Y_2^2.$$

The first-order condition determining Y_1 is

$$\partial(U_1 + U_2)/\partial Y_1 = 1 - 2a_1 Y_1 = 0.$$

Similarly for region 2

$$\partial(U_1 + U_2)/\partial Y_2 = 1 - 2a_2 Y_2 = 0.$$

The equilibrium revenues (superscript b stands for bench-mark) are

$$Y_1^{\mathrm{b}} = 1/2a_1 \tag{2.10}$$

and

$$Y_2^{\mathrm{b}} = 1/2a_2. \tag{2.11}$$

Using equations (2.4) and (2.5) the equilibrium tax rates are

$$x_1^{\mathrm{b}} = 1/2 - a_1/2a_1 + Ea_1 \tag{2.12}$$

and

$$x_2^{\mathrm{b}} = 1/2 - a_2/2a_1 + Ea_2. \tag{2.13}$$

Equations (2.10)–(2.13) will be compared to the results in the next section, where the commitment and no commitment regimes with non-cooperative regions are discussed. It will be seen that the bench-mark case discussed in this section yields the highest revenue levels and lower tax rates compared to the two cases in the following section. In terms of the center's utility level, this outcome is the first best. The reasons are two-fold. First, by maximizing the joint utility function, both regions internalize the benefits of their revenue collection on the other region. Mathematically, equations (2.6) and (2.7) are taken into account by the two regions' union when making decisions about revenue collection. For each region, the incentive to collect a high level of revenue will be stronger than in the case where the positive externality on other regions – equations (2.6) and (2.7) – is ignored. Secondly, taxation in this system does not create distortions since they are equivalent to lump-sum taxes. This is because *ex post* (after the regions have collected the revenues), a proportional tax on the collected revenue does not affect the tax base. However, this hypothetical set-up is not realistic in the sense that current institutions do not ensure the cooperativeness that is assumed in this section. It is interesting therefore to explore the mechanisms that can produce outcomes close to the first best.

2. NON-COOPERATIVE REGIONS: COMMITMENT VERSUS NO COMMITMENT

This section discusses the realistic case where the two regions are non-cooperative. In this set-up, each region maximizes its individual utility, that is it disregards the benefits of its revenue collection on the other region. In the first subsection we will maintain the timing assumption that the regions move first and the center second, that is the center does not commit to pre-announced tax rates. We will show that because the regions do not internalize the positive externality of their revenue collections on other region, each will collect less revenues compared to the bench-mark case. In the second subsection, we will consider the case where the timing of the game is reversed so that the center moves first and the regions move second. Under this regime, the center can commit to pre-announced tax rates and, as a result, the regions collect more revenues than in the no commitment case. Although revenue collection in the commitment case is still lower than in the case of cooperative regions, it is closer to the bench-mark than the no commitment case.

The center cannot commit

Assume that the timing and the players' choice variables in the game are the same as in the previous section. However, instead of assuming the two regions maximize a joint utility function, we assume here that each maximizes its individual utility, that is the two regions are non-cooperative. Between the two regions, they play a Cournot game in which each player does not have the information about the other player's move, as they move simultaneously.

The equilibrium is now defined by three conditions: the center responds optimally given the regions' collected revenues; region 1 optimizes given the center's reaction function and region 2's choice of revenue collection; and region 2 optimizes given the center's reaction function and region 1's choice of revenue collection.

To solve for the equilibrium, we again use backward induction. In other words, we first solve for the center's optimal choices given Y_1 and Y_2 (the center's reaction functions to regions' choices) and then solve for the regions' optimal Y_1 and Y_2 given the center's reaction functions. It is easy to see that the center's reaction functions to the regions' revenues are the same as equations (2.4) and (2.5).

Region 1 now maximizes its individual utility:

$$U_1 = (1 - x_1(Y_1, Y_2))Y_1 - a_1 Y_1^2$$

where $x_1(Y_1,Y_2)$ is defined by equation (2.4) and Y_2 is taken as given. Substituting equation (2.4) into U_1 yields

$$U_1 = Y_1/2 + (Y_2 - E)/2 - a_1 Y_1^2.$$

The first-order condition determining the optimal Y_1 is

$$\partial U_1/\partial Y_1 = 1/2 + \partial Y_2/\partial Y_1 - 2a_1 Y_1 = 0.$$

For region 2, similarly, the first-order condition is

$$\partial U_2/\partial Y_2 = 1/2 + \partial Y_1/\partial Y_2 - 2a_2 Y_2 = 0.$$

Using the Cournot game assumption, $\partial Y_i/\partial Y_j = 0$, where i, $j = 1$, 2, and solving the two first-order conditions simultaneously, the equilibrium under the no commitment regime (superscript nc stands for no commitment) is

$$Y_1^{nc} = 1/4a_1 \tag{2.14}$$

and

$$Y_2^{nc} = 1/4a_2. \tag{2.15}$$

The center substitutes equations (2.14) and (2.15) into equations (2.16) and (2.17) after observing the realized revenues, Y_1^{nc} and Y_2^{nc}, and obtain the no commitment tax rates

$$x_1^{nc} = 1/2 - a_1/2a_2 + 2Ea_1 \tag{2.16}$$

and

$$x_2^{nc} = 1/2 - a_2/2a_1 + 2Ea_2. \tag{2.17}$$

Comparing this equilibrium with the one in the bench-mark yields the following conclusion: Remark: the two regions collect less revenues under the non-cooperative regime than in the bench-mark case. Accordingly, the center has to impose higher tax rates in order to raise its target level of expenditure.

The reason for regions to collect less revenues in the no commitment case, compared to the bench-mark case, is that each region ignores the positive externality of its revenue collection on the other region, that is, equations (2.6) and (2.7). The problem does not arise in the bench-mark case: the union of the two regions acts as one body so that it internalizes the positive externality of one region's revenue collection on the other.

The center can commit

We claim that by altering the timing of the game, that is let the center move first, the regions will have the incentive to collect more revenues than in the case discussed in the previous subsection. In other words, the problem of regions' inadequate tax efforts in the no commitment case is mitigated. We formally analyze the game below.

The game proceeds as follows.

(1) In stage 1, the center announces x_1 and x_2.
(2) In stage 2, after observing x_1 and x_2, region 1 and region 2 choose Y_1 and Y_2 simultaneously.

In the terms that are familiar to many readers, the above assumption says that the center plays a Stackelberg game with the regions and the center moves first (the center is the leader). Under this regime, the regions take the center's offer of tax rates as given when determining their revenues, while the center takes into account the regions' reaction to tax rates and finds the optimal tax rates. As in the previous subsection, the two regions play a Cournot game in which both players do not have the information about the other player's move. When comparing the equilibrium solution to this game with the no commitment game, we will show that the commitment regime yields a high level of total revenue.

The equilibrium is defined by three conditions: the center acts optimally given the regions' reaction functions; region 1 optimizes given the center's announced tax rates and region 2's revenue collection; and region 2 optimizes given the center's announced tax rates and region 1's revenue collection.

We solve for the regions' reaction functions first. Believing that the central government will commit to a constant tax rate, region 1 chooses Y_1 to maximize

$$U_1 = (1 - x_1)Y_1 - a_1 Y_1^2$$

where x_1 is taken as given. This yields the optimal revenue Y_1 under the commitment regime (superscript c stands for commitment):

$$Y_1^c = (1 - x_1^c)/2a_1. \tag{2.18}$$

For region 2, its optimal revenue is

$$Y_2^c = (1 - x_2^c)/2a_2. \tag{2.19}$$

Now return to the center's problem. The center chooses x_1 and x_2 to maximize

$$U_c = \log(1 - x_1)Y_1 + \log(1 - x_2)Y_2 \tag{2.20}$$

subject to the constraint of

$$E = x_1 Y_1 + x_2 Y_2 \tag{2.21}$$

where Y_1 and Y_2 are given by equations (2.18) and (2.19). Substituting equations (2.18) and (2.19) into equation (2.20), we can specify the Lagrange

$$\begin{aligned} L = {} & \log(1 - x_1)(1 - x_1)/2a_1 + \log(1 - x_2)(1 - x_2)/2a_2 \\ & + \lambda\left[x_1(1 - x_1)/2a_1 + x_2(1 - x_2)/2a_2 - E\right]. \end{aligned}$$

Note that $\lambda > 0$. The two first-order conditions are

$$\partial L/x_1 = -2/(1 - x_1) + \lambda(1 - 2x_1)/2a_1 = 0 \tag{2.22}$$

and

$$\partial L/x_2 = -2/(1 - x_2) + \lambda(1 - 2x_2)/2a_2 = 0. \tag{2.23}$$

Equations (2.22) and (2.23) imply that $x_1 < 1/2$ and $x_2 < 1/2$, otherwise $\partial L/x_1 < 0$ and $\partial L/x_2 < 0$. Eliminating λ from the above conditions gives

$$(1 - x_1)(1 - 2x_1)/a_1 = (1 - x_2)(1 - 2x_2)/a_2. \tag{2.24}$$

Substituting equations (2.18) and (2.19) into equation (2.21), the budget constraint can be rewritten as

$$E = (1 - x_1)x_1/2a_1 + (1 - x_2)x_2/2a_2. \tag{2.25}$$

Equations (2.24) and (2.25) implicitly define the optimal tax rates under commitment. Solving these two equations yields a third-order polynomial equation. However, for the purpose of comparing the outcomes of the commitment and no commitment regimes, an explicit solution is not necessary.

Now let us compare this equilibrium with the one in the bench-mark case and the one in the no commitment case discussed in the previous subsection. Comparing equations (2.18) with (2.10) in the bench-mark case, we see that

$$Y_1^c < Y_1^b$$

since $x_1 > 0$. Similarly, we have

$$Y_2^c < Y_2^b$$

since $x_2 > 0$. This states that the commitment regime yields less revenues than the bench-mark case.

Now compare the equilibria under the commitment and the no commitment cases. From equations (2.14) and (2.18) we see that

$$Y_1^{nc} < Y_1^c$$

since $x_1^c < 1/2$. Similarly, we have

$$Y_2^{nc} < Y_2^c$$

since $x_2^c < 1/2$.

Remark: if the central government can commit to fixed tax rates, the regions will collect more revenue than under the no commitment regime. Although the commitment regime is not able to replicate the equilibrium in the bench-mark case, the revenue collection under the commitment regime is closer to that in the bench-mark case.

Comparing this equilibrium with the ones in the bench-mark case and the no commitment case, we can identify the major differences of the regions' tax collection incentive between the three regimes (see Table 2.2). In the bench-mark case, the two regions' union takes into account the positive externality of their revenue collection on other regions, that is $\partial x_i/\partial Y_j < 0$ for $i, j = 1, 2$, and the taxes do not have a distortionary effect on revenues because they are imposed after the revenues have been collected, that is

TABLE 2.2 *Externality and tax distortion in the three systems*

	Bench-mark	*Commitment*	*No commitment*
Whether regions take into account the externality	Yes	Does not exist	No
Tax distortion	No	Yes	Yes

NOTE: 'does not exist' means that no externality issue arises in this system.

$dY_i/dx_i = 0$ because of $Y_i = 1/2a_i$. Under the no commitment regime, each region ignores the externality, since it maximizes its own utility. This ignorance of externality leads to a lower revenue collection. Moreover, the no commitment case involves a tax distortion as regions are tempted to reduce their tax efforts in order to avoid high tax rates.

Under the commitment regime, the externality problem – one region's revenue collection affects the other region's tax rate – disappears since the center pre-commits to fixed tax rates. The elimination of the externality problem increases the regions' tax collection efforts. However, the predetermined tax rates create distortions on the regions' revenue collection: from equations (2.18) and (2.19), that is $Y_i^c = 1 - x_i^c/2a_i$, a positive tax leads to a revenue collection lower than the first best $Y_i = 1/2a_i$. Therefore, when comparing with the no commitment case, two factors should be considered. On one hand, the absence of the externality problem contributes to a higher revenue collection effort in the commitment case, compared to the no commitment case; on the other hand, both systems involve a tax distortion that reduces regions' tax efforts. The aggregated effect of these factors yields higher revenue collection levels in the commitment regime than in the no commitment regime.

3. WHY DOES THE CENTER NOT WANT TO COMMIT: A TIME-INCONSISTENCY EXPLANATION

The commitment regime provides a way to overcome the regions' incentive problem under the no commitment regime. However, the evidence discussed in the introduction suggests that the Chinese central government did not choose to commit. More precisely, it may have tried to commit, as manifested by signing contracts with the regions, but failed to honor the contracts. Why is this the case? Two factors are relevant. First, the institution does not restrict the center from reneging on pre-announced fiscal policy, that is there is no legal restrictions on the center's contract revision. In the terms of Persson and Tabellini (1990), this provides the center with the technology to renege. Second, knowing that the commitment regime yields a higher level of revenue than the no commitment regime, the center still has the incentive not to commit. And this second issue is the topic of this section.

The first subsection shows that the nature of the no commitment problem is that the center's policy is time inconsistent. Suppose the center pre-announces the optimal tax rates (the rates under commitment). After having observed the realized revenues, the center will have the incentive to change the pre-announced tax rates. In Kydland and Prescott's (1977) terms, the optimal policy is time inconsistent. Recognizing this, the promise made by the center becomes incredible to the regions with rational expectation. In the second subsection, a numerical example shows the intuition of this conclusion.

Time inconsistency of the commitment policy

This subsection presents proof of the assertion that the center's optimal policy is time inconsistent if the center is allowed to renege on its pre-announced policy. Suppose that the center announces the optimal tax rates x_1^0 and x_2^0 in the first stage of the game and the regions believe this promise. The regions' optimal responses would be (see equations (2.18) and (2.19))

$$Y_1 = (1 - x_1^0)/2a_1 \tag{2.26}$$

and

$$Y_2 = (1 - x_2^0)/2a_2. \tag{2.27}$$

It is easy to see from equation (2.16) that the center maximizes its utility when

$$(1 - x_1^0)Y_1 = (1 - x_2^0)Y_2. \tag{2.28}$$

This says that when equation (2.28) is satisfied, the center will not regret on the pre-announced tax rates x_1^0 and x_2^0, that is these tax rates will be time consistent.

Now let us examine whether the optimal policy – the tax rates under commitment – is time consistent. Recall that in the previous section (pp. 46–7), we showed that the optimal policy is simultaneously defined by equations (2.24) and (2.25). Combining equation (2.24)

$$(1 - x_1^0)(1 - 2x_1^0)/a_1 = (1 - x_2^0)(1 - 2x_2^0)/a_2$$

with equation (2.28), we see that for the optimal policy to be time consistent, the following must be true:

$$x_1^0 = x_2^0. \tag{2.29}$$

Note that equation (2.29) holds if and only if $a_1 = a_2$. Therefore, for any unequal a_1 and a_2, the optimal policy is time inconsistent.

Remark: unless the two regions' cost coefficients of tax collection are the same, the center will find it optimal to renege on the pre-announced optimal tax rates after the regions have responded to the promised rates.

In almost all practical cases $a_1 \neq a_2$ and the center's optimal policy is not time consistent. What is the consistent policy? From the derivation of no commitment solution shown earlier (p. 45), one can easily see that the no commitment tax rates

$$x_1^{nc} = 1/2 - a_1/2a_2 + 2Ea_1$$

and

$$x_2^{nc} = 1/2 - a_2/2a_1 + 2Ea_2$$

are time consistent, because they satisfy equation (2.28).

A numerical example of the time-inconsistency problem

A numerical example based on the models in the first two sections is given in Tables 2.3 and 2.4. These tables show the equilibrium outcomes of the two regions' tax rates, revenue collections, and the regions' and the center's utility levels for three different cases. These three cases are the no commitment regime, as discussed on pp. 43–5, the commitment regime, as discussed on pp. 45–8 and the case where regions mistakenly believe that the center commits to the pre-announced optimal tax rates and the center re-optimizes after observing the regions' revenue collections. Table 2.3 presents the results when region 1 has a lower cost of revenue collection than region 2, that is $a_1 < a_2$. Table 2.4 presents the results when the two regions' costs of revenue collection are the same.

First look at Table 2.3, the case of $a_1 < a_2$. Under the no commitment regime (column 1), the regions move first and the center moves second. After the regions have collected their revenues, they no longer have the opportunity to react to the center's tax rates, that is *ex post* Ys are perfectly inelastic. With no concern about the negative reaction from the regions, the center's only criterion for choosing optimal tax rates is to equalize after-tax revenues. We therefore observe $(1 - x_1)Y_1 = (1 - x_2)Y_2$ in column 1.

Under the commitment regime (column 2), the center moves first and the regions move second. The center's choice of tax rates has a negative impact on the regions' revenue collection, that is *Y*s are elastic to the *x*s. When calculating the optimal tax rates, the center takes into account the regions' reaction. Since the center moves first, it has to consider the two effects of its tax rates: the distribution effect, that is the impact of tax rates on the distribution of after-tax revenues of the two regions and the efficiency effect, that is the impact of the tax rates on the two regions' revenue collection – $Y_i = (1 - x_i)/2a_i$ for $i = 1, 2$. (While in the no commitment case, there is no efficiency effect.) A set of tax rates that equalizes after-tax revenues across regions involves a high tax rate on the low cost (low a_i) region and a low tax rate on the high cost (high a_i) region. However, this set of equalizing tax rates has a negative impact on total revenue collection since the low cost region, or the more productive region (in the sense that given the same tax rate it collects more revenue than the other region[7]), is punished by a high tax rate. Taking into account the efficiency effect, the center will offer the low cost region a relatively more favorable policy (that is, a lower tax rate relative to the case where the center cares only about equity). Comparing column 1 with column 2 in Table 2.3, we can see that x_1 in the commitment case is less than x_1 in the no commitment case which yields a complete equalization of after-tax revenues. The table also confirms that the two regions' revenue collections (Y_1 and Y_2) are higher in the commitment case than in the no commitment case.

If the center is allowed to renege on its pre-announced optimal tax rates (the 'commitment' tax rates) after the regions have collected the 'commitment' revenues, the center will consider only the distribution effect, not the efficiency effect. This is because *ex post* the regions' revenue collections are no longer responsive to tax rates. Since the low cost region is left with a higher after-tax revenue than the high cost region in the commitment regime, the center will now raise the low cost region's tax rate and lower the high cost region's tax rate in order to achieve equity. Column 3 shows the case where the center is allowed to re-optimize the tax rates, given that the regions collect the 'commitment' revenues. From column 3, one can see that x_1 is higher than that in column 2 (in the commitment case) and x_2 is lower than that in column 2. Comparing the center's utility in columns 2 and 3, it is clear that the tax rate revision increases the center's utility level. This suggests that *ex post* equalization is the center's motivation of reneging on its pre-announced tax rates.

Table 2.4 is an example of the case where the two regions have identical costs of revenue collection. By symmetry, the optimal policy under the commitment regime yields equal tax rates and revenue collections for the two

TABLE 2.3 *Simulation results of three regimes: heterogeneous regions ($a_1 = 0.100$, $a_2 = 0.200$, $E = 1.5$)*

Regime	*No commitment*	*Commitment*	*Reoptimization*
x_1	0.550	0.324	0.427
x_2	0.100	0.203	0.029
Y_1	2.500	3.378	3.378
Y_2	1.250	1.994	1.994
Y_1+Y_2	3.750	5.371	5.371
$(1-x_1)Y_1$	1.125	2.282	1.936
$(1-x_2)Y_2$	1.125	1.590	1.936
U_1	0.500	1.141	0.795
U_2	0.813	0.795	1.141
U_c	0.102	0.560	0.574

No commitment: figures are calculated using equations (2.14)–(2.17). Commitment: x_1 and x_2 are calculated using equations (2.24) and (2.25); Y_1 and Y_2 are calculated using equations (2.18) and (2.19). Reoptimization: Y_1 and Y_2 are taken from the commitment solution. Given these *Y*s, we reoptimize U_c to find the *ex post* optimal tax rates x_1 and x_2.

TABLE 2.4 *Simulation results of three regimes: homogeneous regions ($a_1 = 0.100$, $a_2 = 0.100$, $E = 1.5$)*

Regime	*No commitment*	*Commitment*	*Reoptimization*
x_1	0.300	0.184	0.184
x_2	0.300	0.184	0.184
Y_1	2.500	4.081	4.081
Y_2	2.500	4.081	4.081
Y_1+Y_2	5.000	8.162	8.162
$(1-x_1)Y_1$	1.750	3.330	3.331
$(1-x_2)Y_2$	1.750	3.331	3.331
U_1	1.125	1.665	1.666
U_2	1.125	1.665	1.665
U_c	0.486	1.045	1.045

No commitment: figures are calculated using equations (2.14)–(2.17). Commitment: x_1 and x_2 are calculated using equations (2.24) and (2.25); Y_1 and Y_2 are calculated using equations (2.18) and (2.19). Reoptimization: Y_1 and Y_2 are taken from the commitment solution. Given these *Y*s, we reoptimize U_c to find the *ex post* optimal tax rates x_1 and x_2.

regions (column 2). In other words, equity and efficiency are achieved simultaneously in the commitment case. Therefore, the center does not have the incentive to revise the tax rates even if it has the technology to do so. The reoptimization solution (column 3) is identical to the commitment solution.

4. ADDRESSING THE TIME-INCONSISTENCY PROBLEM

The literature on the time-inconsistency problem has provided us with some valuable insights as to how to tackle the issue of the government's lack of credibility. When discussing the monetary games between the government and the wage-setters, Barro and Gordon (1983), Rogoff (1985), and Canzoneri (1985), among others, suggested that the legislation of a simple monetary rule, the government's reputation effect, and the appointment of a 'conservative' central banker may help address or mitigate the government's credibility problem. Along the line of these arguments, this section discusses several possible institutional changes to address the Chinese central government's time-inconsistency problem in its fiscal relations with the regions. The first proposed change, discussed in the first subsection, is to restrict legally the center's authority to change revenue-sharing methods with the regions. The second approach, discussed in the second subsection, calls for a long-term thinking central government. The third subsection illustrates that an egalitarian center can improve the regions' incentive to collect revenues by appointing a 'less egalitarian' finance minister. The final subsection argues that by institutionalizing a 'benevolent' central government – which maximizes the sum of regions' utility rather than equalizes after-tax revenues – the outcome under no commitment can be improved. The limitations of all these schemes are also discussed in the subsections.

Restricting the center's discretion in changing intergovernmental fiscal relations

In the current Chinese system, the division of taxing powers and expenditure responsibilities is largely determined at the discretion of the central government, rather than through legislation of the National People's Congress. This allows the central government to frequently change the 'rules of the game' in its own interests. As we have shown in previous sections, such a discretion causes serious incentive problems. Regions that rationally expect the center's future policy changes tend to reduce their revenue collection efforts. To address this incentive problem, a mechanism that can restrict the center's discretionary policy change must be established. A law that defines the center's and the regions' rights and obligations regarding fiscal revenue and expenditure and certain institutions that can oversee the implementation of relevant laws, will be a solution of this type. More specifically, the law should stipulate that any major change of the central–local revenue-sharing formula must go through certain legal procedures, perhaps the approval from the Congress. This will substantially increase the center's cost of policy revision and, hence, eliminate or mitigate the regions' incentive problem.

In this regard, the experience of many federalist countries and even unitary countries such as Japan provides some valuable lessons. In these countries, the division of taxing power and expenditure responsibilities – and, consequently, local fiscal autonomy – has been guaranteed by the Constitution. A chapter of Japan's post-war Constitution states that regulations concerning organization and operations of local public entities should be fixed by laws in accordance with the principle of local autonomy. Because the constitution cannot specify the details of tax and expenditure assignment to each level of government, this task is left for laws on local taxation, such as the Local Finance Law and Local Tax Law. Among many laws regarding fiscal issues, these two laws have been the main pieces that govern the division of taxing power and expenditure responsibilities between the central and local governments. They have also protected local governments from arbitrary burden-sharing requirements that might otherwise have been imposed by the central bureaucracy. In addition, the rules specified in the laws have helped to develop local governments' trust in the national government. Once the tax and expenditure assignments are written in law, they cannot be easily altered at the central government's discretion. This provides the central government with a device of commitment to a stable intergovernmental fiscal relation. With such a commitment, the local governments' tax and expenditure behavior becomes more predictable, as the strategic reactions to possible changes in the central–local fiscal arrangement are reduced or eliminated. A predictable pattern of local fiscal behavior will, in turn, contribute to the central government's capacity of macroeconomic management.

Also important is the establishment of institutions that can protect the interest of the local governments and counteract the strong tendency of central government agencies to appropriate fiscal powers from the regions, as a legal definition of the central–local fiscal relations is often not sufficient to guarantee its implementation. In Japan, this task is performed successfully by the Ministry of Autonomy. Individual local governments, although of a large number, generally do not have the incentive and the capacity to seek the protection of local autonomy. This problem is mitigated once the Ministry of Autonomy, whose major task is to protect local autonomy, is instituted.[8]

The central government's reputation effect

In the one-shot game discussed in earlier sections, the regions can be certain (except in the case of homogenous regions) that the center has the *ex post* incentive to renege on its pre-announced tax rates. In this model, the center does not have to consider the intertemporal effect of its lack of commitment, since it maximizes only one period utility. In this subsection, we extend this one-shot game to one which can be repeated an infinite number of times and show that when the center is sufficiently concerned with its reputation, the consideration itself may serve as a device for commitment.

We retain the no commitment assumption: in any given period the center chooses the tax rates after observing the revenues collected by the regions. In the infinite period framework, the center's strategy specifies a sequence of tax rates, $\{x_1, x_2\}$, one set for each period. Similarly, region i's strategy specifies a sequence of revenues, Y_i, one for each period, where $i = 1$ and 2.

Let t denote the period, where $t \in \{0, 1, 2, \ldots, \infty\}$. For each t, the outcome of the $t - 1$ proceeding plays of the one-shot game are observed before the tth period begins. Each player's payoff is the present value of the player's payoffs from the infinite sequence of the one-shot game. The discount value, δ, is constant and shared by the center and the two regions.

A subgame perfect equilibrium of this repeated game is a set of the center's tax rates and a collection of the regions' strategies, one for each region that satisfy two conditions:

(1) Starting from any history of the game, given the center's policy rule (tax rates as functions of the collected revenues) and the equilibrium revenue of the other region, the revenue is optimal for each region.
(2) Given the equilibrium revenues, the tax rates are optimal for the center.

It is easy to see that the no commitment equilibrium in the one-shot game (see pp. 43–5) is also an equilibrium of the repeated game. Recall the timing assumption of the no commitment game that in each period (each stage game) the regions move first. Suppose that in every period regions collect $Y^{\text{nc}} = \{Y_1^{\text{nc}}, Y_2^{\text{nc}}\}$ as in the one-shot game, irrespective of the aggregate history. This strategy is clearly optimal for the regions if, in every period, the center sets tax rates at x^{nc} after observing the collected revenues. It is also true that, given the regions choosing Y^{nc} in every period, it is optimal for the center in every period to choose $x^{\text{nc}} = \{x_1^{\text{nc}}, x_2^{\text{nc}}\}$ as in the one-shot game.

Hence, both conditions for an equilibrium are satisfied and this constitutes the equilibrium of 'no commitment forever'.

'No commitment forever' is not the only equilibrium in this repeated game. It is possible to find some other strategy that can support an equilibrium with higher revenues and lower tax rates. Below we investigate such an equilibrium that involves trigger strategies. Denote the vector of the regions' revenues in period t with Y_t and the vector of the center's tax rates on the two regions with x_t. Consider the following trigger strategy of the regions

$$\begin{aligned} \text{For } t = 0, Y_0 &= Y^{\mathrm{c}} \\ \text{For } t > 0, Y_t &= Y^{\mathrm{c}} \quad \text{if } x_{t-1} = x^{\mathrm{c}} \\ &= Y^{\mathrm{nc}} \text{ otherwise} \end{aligned} \tag{2.30}$$

and the trigger strategy of the center

$$\begin{aligned} \text{For } t = 0, x_0 &= x^{\mathrm{c}} \\ \text{For } t > 0, x_t &= x^{\mathrm{c}} \quad \text{if } Y_t = Y^{\mathrm{c}} \\ &= x^{\mathrm{nc}} \text{ otherwise} \end{aligned} \tag{2.31}$$

where $t = 1, 2, \ldots \infty$. The regions start by choosing revenues on the expectation of the center's commitment. If this expectation is fulfilled they continue to choose revenues of the commitment equilibrium, Y^{c}. Otherwise, they revise their expectation into one of no commitment and choose the no commitment revenue, Y^{nc}. In other words, as long as the center sticks to the commitment tax rates, the regions behave cooperatively. Once the center deviates from the commitment equilibrium, the regions shift to a non-cooperative regime by choosing the no commitment revenue level in the next period and thereafter. Similarly, the center sets x^{c}, the commitment tax rates, provided that the current period revenue collections are cooperative, that is $Y = Y^{\mathrm{c}}$; otherwise, the center shifts to the non-cooperative regime by choosing x^{nc}.

Suppose we start from the initial period when $Y_0 = Y^{\mathrm{c}}$ and $x_0 = x^{\mathrm{c}}$. The following formula characterizes the condition under which the center would not deviate away from the cooperative strategy:

$$U(Y^{\mathrm{c}}, x^{\mathrm{r}}) - U(Y^{\mathrm{c}}, x^{\mathrm{c}}) \leq [U(Y^{\mathrm{c}}, x^{\mathrm{c}}) - U(Y^{\mathrm{nc}}, x^{\mathrm{nc}})]\delta/(1-\delta) \tag{2.32}$$

where $U(Y^{\mathrm{c}}, x^{\mathrm{r}}) - U(Y^{\mathrm{c}}, x^{\mathrm{c}})$ is the center's gain in utility if it shifts from a cooperative to a non-cooperative regime in period 1 (we only consider a shift in period 1 since the incentive of such a shift in any period t is the same as the incentive of a shift in period 1). x^{r} is the set of the center's reoptimized tax rates (that is, the 'cheating' strategy) given the regions' commitment revenues. It is defined as

$$x^{r} = \underset{\{x_1, x_2\}}{\operatorname{argmax}} \log(1 - x_1)Y_1 + \log(1 - x_2)Y_2$$

subject to

$$\text{(i) } E = x_1 Y_1 + x_2 Y_2$$
$$\text{(ii) } Y_i = (1 - x_i)/2a_i$$

where $i = 1, 2$. Recall that constraint (ii) gives the regions' optimal revenues in the commitment equilibrium (see equations (2.18) and (2.19)).

The center will lose utility $U(Y^{c}, x^{c}) - U(Y^{nc}, x^{nc})$ in each period starting from period 2 if it chooses to deviate in the first period. The right-hand side of equation (2.32) gives the sum of the discounted value of these losses. From the discussion in the previous section we know that

$$U(Y^{c}, x^{r}) - U(Y^{c}, x^{c}) \geq 0 \tag{2.33}$$

and

$$U(Y^{c}, x^{c}) - U(Y^{nc}, x^{nc}) \geq 0. \tag{2.34}$$

Equation (2.33) states that the center has the *ex post* incentive to break the commitment in a one-shot game. Equation (2.34) states that a commitment regime is superior to a no commitment regime in terms of the center's utility level. This is obvious since under the commitment regime the center plays the game as a Stackelberg leader.

In period 0 and any other period in which all the proceeding outcomes have been $\{Y^{c}, x^{c}\}$, the center's optimal action is – given that the regions have adopted the trigger strategy – Y^{c} if and only if equation (2.32) is satisfied. Equation (2.32) therefore states that the center will shift from a commitment regime to a no commitment regime as long as it finds that, in present value terms, the total losses of reneging on the commitment tax rates is smaller than the associated gain.

Remark: holding all other parameters constant, if δ is sufficiently close to one, the center will find it optimal to stick to the commitment strategy and, consequently, 'commitment forever' will be the equilibrium outcome.[9]

Intuitively, a central government that values its reputation (δ is high) will be likely to keep its commitment, since any deviation from the commitment path will cause significant future losses in present value terms. In contrast, if the center is short-sighted (with a small δ), it will likely deviate from the commitment equilibrium in the first period. It is reasonable to argue that the time discount factor (δ) of the current Chinese central government is too small to support a cooperative equilibrium, since it has failed to commit to the fiscal contracts. An immediate implication from this analysis is that freeing the central fiscal authorities from the short-term equalization objective and emphasizing the medium and long-term objective of increasing regions' incentive to collect revenues will help mitigate the time-inconsistency problem.

A number of technical difficulties make the reputation approach less appealing than it appears to be. First, the reputation effect will work only in the case of an infinitely repeated game, which is not a realistic assumption since the government has a limited term of office. With a finitely repeated game, the 'commitment forever' equilibrium is not subgame perfect. Namely, if one works backwards to solve the game, the only equilibrium is 'no commitment forever' (see Persson and Tabellini, 1990; Fudenberg and Tirole, 1991). Second, the commitment regime is difficult to sustain in the presence of uncertainties. For example, there can be an *ex post* shock to the expenditure target which is observable to the center but never to the regions. The center has to adjust tax rates to maintain the budget balance. The regions are not able to perfectly verify the center's commitment – whether its tax rate adjustment is due to shock financing or due to redistribution. Even if the center does commit not to adjust tax rates for redistributive purposes, the regions may still choose to shift to a non-cooperative regime when observing a large *ex post* change in tax rates.

Appointing a 'less egalitarian' finance minister

In earlier sections, we assumed that the center is perfectly egalitarian: it gives equal weights to the two regions' after-tax revenues in the utility function. This subsection argues that it may be desirable for such a center to appoint a finance minister – who decides the revenue-sharing methods with the regions – who is less egalitarian than itself. By doing so the regions as a whole will collect more revenues and the center's utility will be higher. This result is analogous to that in the monetary literature which suggests that appointing a central banker who is more 'conservative' (inflation hating) than the society may reduce the government's incentive to create surprise inflation and, hence, increase social welfare.

To formalize this argument, we suppose that the center can appoint a finance minister who chooses x_1 and x_2 to maximize the following objective:

$$U_c = \phi \log(1 - x_1)Y_1 + \log(1 - x_2)Y_2 \tag{2.35}$$

subject to

$$x_1 Y_1 + x_2 Y_2 \geq E \tag{2.36}$$

where ϕ, which is positive but can be different from 1, is the weight that the finance minister gives to region 1. Under the no commitment regime, the center chooses the following optimal tax rates:

$$x_1 = \phi/(1+\phi) - \phi Y_1/(1+\phi)Y_2 + \phi E/(1+\phi)Y_1 \tag{2.37}$$

and

$$x_2 = 1/(1+\phi) - Y_2/(1+\phi)Y_1 + E/(1+\phi)Y_2. \tag{2.38}$$

Taking into account these reaction functions, the regions, with the same objectives as before, can solve for their optimal revenues. Region 1's problem is to choose Y_1 to maximize

$$U_1 = (1 - x_1)Y_1 - a_1 Y_1^2 \tag{2.39}$$

where x_1 is defined by equation (2.37). Substituting equation (2.37) into equation (2.39) and solving the first-order condition yield the optimal revenue for region 1:

$$Y_1^{nc} = \phi/[2(1+\phi)a_1]. \tag{2.40}$$

Similarly, the optimal revenue for region 2 is

$$Y_2^{nc} = 1/[2(1+\phi)a_2]. \tag{2.41}$$

Substituting equations (2.40) and (2.41) back to equations (2.37) and (2.38) gives the tax rates in equilibrium:

$$x_1^{nc} = \phi/(1+\phi) - a_1/(1+\phi)a_2 + 2a_1 E \tag{2.42}$$

and

$$x_2^{nc} = 1/(1+\phi) - \phi a_2/(1+\phi)a_1 + 2a_2 E. \tag{2.43}$$

Without loss of generality, let us assume that $a_1 < a_2$. We call region 1 a more productive region since it has a lower revenue collection cost than region 2. Recall that the center can appoint a finance minister who attaches a weight ϕ to region 1. What is the optimal ϕ that the perfectly egalitarian center should choose to maximize its own utility? By substituting equations (2.40)–(2.43) to the center's utility function,

$U_c = \log(1 - x_1)Y_1 + \log(1 - x_2)Y_2$, we have the center's indirect utility function expressed by a_1, a_2, and ϕ:

$$\begin{aligned} U_c &= \log[1/(1+\phi) + a_1/(1+\phi)a_2 - 2a_1E]\phi/[2(1+\phi)a_1] \\ &\quad + \log[\phi/(1+\phi) + \phi a_2/(1+\phi)a_1 - 2a_2E]/[2(1+\phi)a_2] \\ &= \log[\phi/(1+\phi)^2] + 2\log[\phi/(1+\phi)2a_1 + 1/(1+\phi)2a_2 - E]. \end{aligned}$$

Taking first derivative of U_c with respect to ϕ yields

$$\begin{aligned} \partial U_c/\partial\phi = (1-\phi)/[(1+\phi)\phi] + (a_2 - a_1)/[(a_2\phi - a_1 \\ + (1+\phi)2a_1a_2E)^2(1+\phi)]. \end{aligned}$$

Evaluating this derivative at $\phi = 1$, we see that

$$\partial U_c/\partial\phi \mid_{\phi=1} = (a_2 - a_1)/[2(a_2 - a_1 + 4a_1a_2E)^2] > 0$$

since $a_2 > a_1$.

Let us call the central government that equally values the two regions' after-tax revenues the 'egalitarian center' and the finance minister with $\phi > 1$ 'less' egalitarian. The above analysis suggests the following remark.

Remark: it is optimal for the egalitarian center ($\phi = 1$) to appoint a finance minister who is 'less' egalitarian ($\phi > 1$) than itself.

We can compare the total revenue collected under the egalitarian center and that under the less egalitarian minister. From equations (2.16) and (2.17), the total revenue collected under the egalitarian center is

$$1/4a_1 + 1/4a_2 \tag{2.44}$$

while the total revenue collected under the less egalitarian minister is

$$\phi/[2(1+\phi)a_1] + 1/[2(1+\phi)a_2]. \tag{2.45}$$

It is easy to show that

$$\phi/[2(1+\phi)a_1] + 1/[2(1+\phi)a_2] > 1/4a_1 + 1/4a_2$$

since $a_1 < a_2$ and $\phi > 1$.

Remark: under the less egalitarian finance minister, the total revenue in equilibrium is higher than that under the egalitarian center.

With a higher weight attached to the low cost region 1, the finance minister will charge region 1 a lower tax and charge region 2 a higher tax than those if the weights to both regions are the same. Intuitively, as the policy maker becomes 'less egalitarian', the low cost or more productive region obtains stronger incentive to collect revenues because the policy applied to this region becomes more favorable. The revenue collection of this region therefore increases (the first term in equation (2.45) increases as ϕ increases). Although the high cost region's revenue collection will fall (the second term in equation (2.45) falls as ϕ decreases) because its policy becomes less favorable, the increase in the low cost region's revenue collection outweighs the reduction in the high cost region's revenue collection. Overall, the total revenue rises.

In the equilibrium of a less egalitarian finance minister, region 1's after-tax revenue is higher than that of region 2. This suggests that although introducing a less egalitarian finance minister does not solve the center's time-inconsistency problem *per se*, it mitigates the negative impact of the center's time-inconsistent behavior on regions' revenue collection by not completely equalizing the two regions' after-tax revenues. Simply speaking, this mechanism induces higher tax efforts from the productive region by offering it a higher rate of revenue retention than that to the other region.

A benevolent central government

In earlier sections we assumed that the central government maximizes the sum of the log of two regions' after-tax revenues, rather than the regions' utilities. This assumption is supported by the empirical evidence presented in the first section (pp. 39–40). With this assumption we generated the time-inconsistency problem as manifested by the Chinese central government's repeated revision of central–local revenue-sharing methods.

This subsection argues that if the central government cares about the regions' utilities instead of their after-tax revenues, the outcome will be very different. Specifically, under the no commitment regime, the first best outcome can be achieved by such a change in the center's objective.

We label the central government that cares about the two regions' utilities as 'a benevolent' center. The objective of such a center can be written as

$$U_c = \log[(1-x_1)Y_1 - a_1 Y_1^2] + \log[(1-x_2)Y_2 - a_2 Y_2^2]. \qquad (2.46)$$

The center faces the same budget constraint as before:

$$x_1 Y_1 + x_2 Y_2 \geq E.$$

Assuming the second-order conditions are satisfied, and following the same procedures as used in the second section (pp. 43–8), the center's optimal tax rates are

$$x_1 = [Y_1 - Y_2 + E + a_2 Y_2^2 - a_1 Y_1^2]/2Y_1 \tag{2.47}$$

and

$$x_2 = [Y_2 - Y_1 + E + a_1 Y_1^2 - a_2 Y_2^2]/2Y_2. \tag{2.48}$$

Assume realistically that regions are non-cooperative. Their problems are the same as those in the third section. They take into account the reaction functions defined by equations (2.47) and (2.48) and choose their optimal levels of revenue collection. Region 1 chooses Y_1 to maximize the following objective

$$U_1 = (1 - x_1)Y_1 - a_1 Y_1^2 \tag{2.49}$$

where x_1 is defined by equation (2.47). Substituting equations (2.47) into (2.49) yields

$$\begin{aligned} U_1 &= (1 - x_1)Y_1 - a_1 Y_1^2 \\ &= [Y_1 + Y_2 - E + a_1 Y_1^2 - a_2 Y_2^2]Y_1/2Y_1 - a_1 Y_1^2 \\ &= Y_1/2 - a_1 Y_1^2/2 + (Y_2 - E + a_2 Y_2^2)/2. \end{aligned}$$

Solving the first-order condition yields the optimal level of revenue for region 1:

$$Y_1 = 1/2a_1. \tag{2.50}$$

Similarly, region 2's optimal revenue is

$$Y_2 = 1/2a_2. \tag{2.51}$$

Substituting equations (2.50) and (2.51) into equations (2.47) and (2.48) we obtain the equilibrium tax rates

$$x_1 = 1/4 - a_1/4a_2 + a_1 E \tag{2.52}$$

and

$$x_2 = 1/4 - a_2/4a_1 + a_2 E. \tag{2.53}$$

We can show that this outcome (equations (2.50)–(2.53)) is equivalent to that derived from a system in which the center uses lump sum taxes. In the lump sum tax system, the center's problem is to select two lump sum taxes to maximize

$$U_c = \log[Y_1 - T_1 - a_1Y_1^2] + \log[Y_2 - T_2 - a_2Y_2^2]$$

subject to

$$T_1 + T_2 \geq E$$

where T_1 and T_2 are lump sum taxes imposed on regions 1 and 2, respectively. Region 1's problem is to select Y_1 to maximize

$$U_1 = Y_1 - T_1 - a_1Y_1^2$$

and region 2's problem is to select Y_2 to maximize

$$U_2 = Y_2 - T_2 - a_2Y_2^2.$$

It is obvious that since the lump sum tax system yields no distortion, the equilibrium is the first best. Therefore the system of a benevolent center also yields the first best outcome. In the system of a benevolent center, the center's lack of commitment (the fact that the center determines taxes after the regions have collected the revenues) does not lead to the regions' incentive problem as in the system of an egalitarian center. The intuition is that with a benevolent center there is no interregional externality problem (a positive or negative impact of one region's action on the other region). The reason is as follows. The rational regions know for sure that *ex post* the center will equalize the utility levels (rather than after-tax revenues) across the two regions, since the first-order condition from maximizing equation (2.46) gives

$$(1 - x_1)Y_1 - a_1Y_1^2 = (1 - x_2)Y_2 - a_2Y_2^2.$$

For region 1, the Y_1 that maximizes its own utility also maximizes the utility of region 2; the same also applies to region 2. In other words, it is impossible for any region to manipulate its revenue collection to increase its own utility at the expense of the other region. In the absence of the interregional externality problem, the outcome is necessarily better than that in the case where each region ignores the positive externality of its own action on other regions.

This result suggests that it would be desirable for the Chinese central government to consider the overall welfare level of the regions rather than focus solely on revenue redistribution when formulating its central–local revenue-sharing methods. The socialist tradition of equalizing per capita after-tax revenues across agents (in our model, it is the equalization across provinces; at the firm level, this tradition yields the equalization of per capita after-tax profits across enterprises) should be down-played in future policy making. Instead, criteria based on some more general measures of welfare level that the regions cannot easily manipulate (such as GNP level, degree of urbanization, school enrolment rate, infant mortality rate, and so on) should be emphasized in achieving the equity objective.

5. THE EFFECT OF THE 1990 PROPOSAL – INCREASING THE CENTER'S SHARE IN TOTAL REVENUE

Since 1980 when the central–local fiscal contract system was adopted, there has been a continued decline of the revenue : GNP ratio and the center has found it increasingly difficult to meet its expenditure targets. Many argued that this was because that the center was too weak in bargaining with the regions. To mitigate the increasing financial difficulties of the central government, in 1990, a proposal of raising 'two ratios' was adopted in the country's Eighth Five Year Plan. One of the ratios to be raised is that of the central government revenue to total government revenue.

There is no systematic analysis of the effect of this proposal. In particular, whether increasing the center's share of revenue can ensure an increase of total revenue as well as the revenue going to the center is still an unanswered question. Here we offer an analysis using the framework developed in the previous sections and show that the 1990 proposal may not be able to solve the problem of the declining revenue : GNP ratio if the discretionary nature of the revenue-sharing system is unchanged.

We consider only the realistic case, that is the no commitment regime with non-cooperative regions. The timing, the players, their choice variables, and the definition of equilibrium of the game are the same as in the subsection on 'the center cannot commit'. The only difference between the regime proposed in 1990 and the system described in that subsection is that the center's budget constraint changes from an expenditure target to one that requires a minimum share of central government revenue in the total government revenue. After solving for the equilibrium, we will use a comparative static analysis to show the impact of an exogenous increase in the center's minimum share in total revenue on the two regions' revenue collections.

First consider the central government's problem. The central government has the same objective as in the subsection on the 'center cannot commit':

$$U_c = \log(1 - x_1)Y_1 + \log(1 - x_2)Y_2 \tag{2.54}$$

where Y_i is per capita revenue. Unlike the model in the second section which assumes a fixed expenditure target, here we assume that the center's revenue is a fixed proportion of the total revenue, $Y_1 + Y_2$:

$$\tau(Y_1 + Y_2) = x_1 Y_1 + x_2 Y_2 \tag{2.55}$$

where τ is a fixed proportion. We assume $0 < \tau < 1$. Maximizing equation (2.54) subject to equation (2.55) gives the optimal levels of x_1 and x_2 as functions of Y_1, Y_2 and τ:

$$x_1 = (1+\tau)/2 - (1-\tau)Y_2/2Y_1 \tag{2.56}$$

and

$$x_2 = (1+\tau)/2 - (1-\tau)Y_1/2Y_2. \tag{2.57}$$

Given the center's reaction function, the regions' problem are specified and solved in the same way as in the second section. The equilibrium levels of revenue for the two regions are

$$Y_1^{nc} = (1-\tau)/4a_1 \tag{2.58}$$

and

$$Y_2^{nc} = (1-\tau)/4a_2. \tag{2.59}$$

In the equilibrium, each region's revenue is determined by its own cost coefficient and τ. The equilibrium tax rates are

$$x_1^{nc} = (1+\tau)/2 - (1-\tau)a_1/2a_2 \tag{2.60}$$

and

$$x_2^{nc} = (1+\tau)/2 - (1-\tau)a_2/2a_1. \tag{2.61}$$

With equations (2.58)–(2.61), we can easily analyze the effect of an increase in τ on Y_1^{nc}, Y_2^{nc}, x_1^{nc}, x_2^{nc}. It is obvious that

$$dY_1^{nc}/d\tau = -1/4a_1 < 0 \tag{2.62}$$
$$dY_2^{nc}/d\tau = -1/4a_2 < 0 \tag{2.63}$$
$$dx_1^{nc}/d\tau = 1/2 + a_1/2a_2 > 0 \tag{2.64}$$

and

$$dx_2^{nc}/d\tau = 1/2 + a_2/2a_1 > 0. \tag{2.65}$$

Remark: an increase in the center's share in total government revenue leads to an increase in tax rates on both regions and a reduction in revenue collection in both regions.

This suggests that the proposal of raising the center's share in total revenues will further dampen the regions' tax efforts under the no commitment regime. Intuitively, this is because a higher τ requires higher tax rates and, with higher tax rates, it is not worthwhile for the regions to collect as much revenue as before. To see this, suppose originally region 1 collects Y_1 when $\tau = \tau_1$. Will this region continue to collect Y_1 as τ increases from τ_1 to τ_2? When $\tau = \tau_2$, the marginal gain of the region's revenue collection, $(1 - x_1)$, is lower because x_1 is higher, but the marginal cost of revenue collection, $2a_1 Y_1$, is unchanged if the region continues to collect Y_1. It is therefore not optimal for this region to stay with Y_1 when $\tau = \tau_2$. By reducing its revenue collection, the region can reduce the marginal cost of revenue collection and improve its welfare.

Now let us look at the effect of a rise in τ on the central government's revenue, $x_1^{nc} Y_1^{nc} + x_2^{nc} Y_2^{nc}$. The derivative of the center's revenue with respect to τ is

$$d(x_1^{nc} Y_1^{nc} + x_2^{nc} Y_2^{nc})/d\tau = (1 - 2\tau)[1/4a_1 + 1/4a_2]. \qquad (2.66)$$

It is positive when $\tau < 1/2$ and negative when $\tau > 1/2$.

Remark: when τ is not too high (less than 1/2), an increase in τ increases the center's revenue. If τ is too high (higher than 1/2), an increase in τ decreases the center's revenue.

The intuition is the same as that of the Laffer curve. An increase in τ has two opposing effects on the center's revenue. On one hand, an increase in τ increases x_1 and x_2 and, consequently, positively contributes to the center's revenue. On the other hand, it reduces Y_1 and Y_2 and consequently negatively affects the center's revenue. If $\tau<1/2$, the positive effect dominates and the center's revenue increases as τ increases. The center's revenue reaches its maximum at $\tau = 1/2$. When $\tau > 1/2$, the negative effect dominates and the center's revenue decreases as τ increases. It is easy to show that at the two extremes of $\tau = 0$ and $\tau = 1$ the center's revenue is zero.

The implication is that an increase in τ cannot be justified without considering the initial level of τ. Nevertheless, there is good reason to believe that an increase in τ will help increase the center's revenue in the current Chinese system since the center's share in China (37 per cent in 1993, see Table 1.2) is lower than most countries in the world (Wang, 1994b).[10] The warning offered here is that the increase in the center's revenue, if achieved, may be at the expense of reducing total revenue collection in the economy.

CONCLUDING REMARKS

China's central–local fiscal relations have been marked by constant changes of the rules of the game. Since 1980, there have been three major revisions in revenue-sharing formulae and for each announced formula there were many adjustments made at the discretion of the center. Although the aim of these changes was to stabilize the center's revenues and to ensure government expenditures in 'key' projects (in infrastructure, agriculture, national defense, and so on), the actual consequence was a rapid decline of both the center's share in the total government revenue and the share of the total government revenue in GNP. This in turn resulted in increasing government deficits, inadequate investment in key projects, and the inability of the central government to carry out reforms that involve significant income redistribution effects.

It seems that the Chinese leadership has not fully understood that tying the hands of the central government can eliminate the strategic response of the localities and can make the outcome more manageable. This is reflected by the fact that China's current central–local relation is not clearly defined by the constitution. Although the center opted to decentralize certain management powers to the localities, it has retained the rights (inherent in the politically centralized system) to change the rules of the game and has frequently reneged on the promises when it thought this was 'necessary'. The localities' rights are not protected by laws, but are defined by the center's interests. While the central government's decision to replace the fiscal contract system with the (incomplete) tax assignment system in 1994 was an important step towards eliminating the localities' incentive to bargain for favorable fiscal treatment, the center has not been willing to give up flexibility in redefining the revenue-sharing formula, the expenditure responsibilities, and the affiliation of state-owned enterprises.[11]

The system in which the center can define central–local fiscal relations at its discretion causes serious incentive problems, an example of which is presented in the central–local fiscal game in this chapter. In this game, regions that rationally expect the center's policy changes tend to reduce their revenue collection efforts. To address this incentive problem, some mechanisms that can restrict the center's discretionary policy change – the 'no commitment' problem that we refer to in the text – must be established. A law that defines the center's and the localities' rights and obligations regarding fiscal revenue and expenditure and certain institutions that can enforce relevant laws will be a solution of this type. More specifically, the law should stipulate that any major change of the central–local revenue-sharing formula must go through certain legal procedures. This will substantially increase the center's cost of policy revision and, hence, reduce its incentive to do so.

In addition to imposing legal restrictions on the center's discretionary policy, three alternative ways to address the center's no commitment problem are discussed in this chapter. First, a center that is sufficiently concerned about its reputation may have less incentive to renege on pre-announced policies. This suggests that freeing the center from short-term adjustment objectives and emphasizing more on the medium and long-term objective of increasing regions' incentive to collect revenues will help mitigate the time-inconsistency problem. Second, if the egalitarian (as defined in the text) center can appoint a less egalitarian finance minister (who attaches a higher weight to the productive region and a lower weight to the less productive region; 'productive' as defined in the subsection. Appointing a 'less egalitarian' finance minister) to design and implement the central–local revenue-sharing policy, the outcome will be better even without changing the no commitment system. Third, if the center can base its revenue redistribution on some general measures of regions' welfare rather than after-tax revenues, the outcome may also improve.

3 Central Government Transfer Under a Soft Budget Constraint

INTRODUCTION

As discussed in Chapter 1, the Chinese central government has experienced increasing difficulties in balancing its budget over the past decade. Two obvious factors are responsible for this problem. First, the current central–local revenue-sharing methods provide the localities with adverse incentives for revenue collection (see Chapter 2). Second, the central fiscal authority has found it very difficult to control the expenditure expansion. The excessive expenditure expansion has to do with two institutional features of the central government's budget formation process: first, the decision making of most investment projects financed by the budget is strongly influenced by lobbying of the local governments and, second, the budget constraint is soft (that is, the budget approved by the Congress is not strictly adhered to); additional expenditures (exceeding the officially approved budget) that accommodate regions' lobbying are common.[1]

This chapter analyzes the process of expenditure target formation and identifies the source of the excessive expenditure expansion in the framework of a central–local game, based on the model developed in Chapter 2. Specifically, we will look at how the center's total expenditure is affected by regions' lobbying. It is shown that under a system of soft budget constraint – a system that does not prohibit regions' lobbying for transfers from the center after the budget is officially approved by the Congress – the regions make too much lobbying efforts and result in excessive expenditure expansion, compared to the system of hard budget constraint in which such *ex post* lobbying is prohibited.

The chapter concludes that the source of the excessive expenditure expansion under the soft budget constraint is a time-inconsistency problem. Before the center's budget plan is determined, the regions find it optimal to commit to a relatively low level of lobbying, since the burden of too much lobbying will be borne by the regions: they have to remit a high share of locally collected revenues to the center if the center's expenditure is high. However, after the center has announced the budget plan, which includes an expenditure target and a revenue-sharing formula (a set of tax rates on the regions, see Chapter 2 for definition) with the regions, the regions will find it optimal to increase their lobbying efforts as the tax rates are no longer responsive to their lobbying efforts. Therefore, the regions' *ex ante* commitment to a low level of lobbying is not credible in the absence of a mechanism that prohibits *ex post* lobbying. Such a system is referred to as one of soft budget constraint. Knowing that the regions are unable to commit, the center sets high tax rates that are compatible with a high level of lobbying in the first place. The equilibrium under a soft budget constraint is therefore a higher level of expenditure, higher tax rates, and a lower welfare level, compared to the case where the regions can restrain themselves from lobbying *ex post* (hard budget constraint).

The modeling of regions' lobbying is in part inspired by the existing work on US interest group lobbying, such as Tullock (1980) and Becker (1983). However, most of the existing works used a simple Cournot game structure, where only two competing lobbying groups' behaviors are explicitly modeled. In this chapter, we develop a model in which both the regions (or the 'interest groups'), which lobby for central government transfers, and the central government, which needs to balance its budget, are explicitly modeled. The co-existence of the interaction among the regions and the interaction between the regions and the center makes the result more interesting. In addition, the identification of the time-inconsistent behavior of the lobbying groups as the source of excessive lobbying efforts is also novel in the literature.

The rest of the chapter is organized as follows. The first four sections assume that the center does not commit to a pre-announced revenue-sharing method. The first section discusses the system of hard budget constraint, in which the regions can only lobby for transfers from the center before the budget plan is formally approved. Any *ex post* lobbying is either prohibited or ineffective. The equilibrium outcome of total lobbying efforts, total expenditure, and revenue collections is derived from this model. The second section looks at a system mimicking the current Chinese system with a soft budget constraint, in which regions can effectively lobby for transfers from the center even after the budget plan is formally approved. The center, when formulating the budget plan, has to take into account the *ex post* lobbying. The equilibrium under this system is derived. The third section compares the equilibrium outcomes of the two systems and indicates that with a soft budget constraint the total expenditure is higher and the regions are worse off. The fourth section asks why in reality the regions do not restrict themselves from *ex post* lobbying, knowing that the soft budget constraint system yields a lower utility level for them. It is argued that the source of this problem is that the regions' policy is time inconsistent. The fifth section uses a general specification of transfer function to derive the Cournot equilibrium of regions' lobbying efforts. It is shown that the effectiveness of regions' lobbying plays an important role in determining total lobbying activities. The sixth section shows that the regions' time-inconsistency problem also exists in the case where the center can commit to a pre-announced revenue-sharing formula. Concluding remarks are given in the last section.

REGIONS' LOBBYING UNDER A HARD BUDGET CONSTRAINT

For analytical purposes, we divide China's central government budget into two components: the expenditures by the central government itself and the transfers to the local governments. The expenditures by the central government include those on central government administration, national defense, repayment of national debts, and some social welfare programs and price subsidies that are uniformly implemented by the central government. This component is relatively stable and predictable. The transfers to the localities consist of various investments in productive projects, including those in the industrial, agricultural, energy, and transportation sectors, as well as subsidies to state-owned enterprises (SOEs) and some and some recurrent expenditures.[2] Decision making related to these transfers is subject to intense lobbying by local governments. A typical form of such lobbying is that a local leader seeks help from a senior leader (sometimes retired, but still influential) in the State Council or the Central Party Committee who is his or her close friend or former supervisor, and the senior leader writes a note to the State Planning Commission suggesting consideration of a project in favor of the locality. Virtually every investment project formulated in the state budget has been backed by these types of 'notes'.[3]

This chapter formalizes the Chinese budget formation process in a game theoretic model. This game is played between the center and two regions, region 1 and region 2. To simplify the analysis, we assume that any investment project (that is, transfer from the center to a region) is located in only one region.

First consider the system of a hard budget constraint. Under this system, once the total expenditure target is determined, further lobbying by the regions is either prohibited or simply does not affect the expenditure plan. The game involves two stages and proceeds as follows.

(1) In the first stage, the two regions choose how much lobbying efforts to make, taking into account the fact that their lobbying efforts will

positively affect the total expenditure and consequently the tax rates (as in Chapter 2, 'tax rate' is defined as the share of locally collected revenue remitted to the center). Also in this stage, the regions determine how much revenue to collect, taking into account that their revenue collections will affect the tax rates. The two regions play a Cournot game in this stage, each taking the other party's action as given when making its own decisions.

(2) In the second stage, observing the regions' lobbying efforts and the collected revenues, the center determines the tax rates on regions 1 and 2.

This game is an extension of the no commitment game discussed in Chapter 2. No commitment means that the center determines the tax rates after observing the regions' revenue collections. We maintain this assumption in this and the following four sections. The same derivation carried out here can be done by assuming a commitment regime in which the center determines fixed tax rates before the regions collect revenues (see the sixth section).

We first look at the center's problem in the second stage. The center chooses the two tax rates (x_1, x_2) to maximize the sum of the logs of the two regions' retained revenues plus the transfers from the center to the two regions:[3]

$$U_c = \log[(1 - x_1)Y_1 + G_1] + \log[(1 - x_2)Y_2 + G_2] \tag{3.1}$$

where Y_1 and Y_2 are per capita revenues of regions 1 and 2, respectively. G_1 and G_2 are the transfers from the center to regions 1 and 2, respectively. In the second stage of the game, Y_1, Y_2, G_1 and G_2 are all given. The budget constraint is

$$x_1 Y_1 + x_2 Y_2 \geq G + G_1 + G_2 \tag{3.2}$$

where G is the expenditure by the central government itself. We assume it is exogenously given.

Solving the center's problem, we obtain the center's optimal tax rates as functions of Y_1, Y_2, G_1, G_2, and G:

$$x_1 = 1/2 - Y_2/2Y_1 + (2G_1 + G)/2Y_1 \tag{3.3}$$

and

$$x_2 = 1/2 - Y_1/2Y_2 + (2G_2 + G)/2Y_2. \tag{3.4}$$

Now let us look at region 1's problem in stage 1. Region 1 chooses an optimal revenue, Y_1, and an optimal level of lobbying effort, L_1, to maximize the sum of its retained revenue and the central government transfer, less the revenue collection cost and lobbying cost:

$$U_1 = (1 - x_1)Y_1 - a_1 Y_1^2 + G_1(L_1) - cL_1 \qquad (3.5)$$

where $(1 - x_1)Y_1$ is the retained revenue and $a_1 Y_1^2$ is the cost of revenue collection (see Chapter 2). L_1 is the lobbying effort and cL_1 is the cost of lobbying in monetary terms, where $c > 0$. We restrict $L_1 \geq 0$. $G_1(L_1)$, the transfer from the center to region 1, is a function of region 1's lobbying effort. We assume the following properties of the transfer function:

$$(A1)\ G_1(0) = 0, \qquad (3.6)$$

that is, zero own lobbying effort yields zero transfer;

$$(A2)\ \partial G_1/\partial L_1 > 0, \qquad (3.7)$$

that is, transfer is an increasing function of own lobbying effort. Similar properties apply to G_2, the transfer function for region 2.[4]

Moving first in the game, region 1 takes into account the center's reaction functions given by equations (3.3) and (3.4). Substituting equations (3.3) and (3.4) into equation (3.5) the region's objective function becomes

$$\begin{aligned} U_1 &= (Y_1/2 + Y_2/2 - G_1 - G/2) - a_1 Y_1^2 + G_1(L_1) - cL_1 \\ &= Y_1/2 + Y_2/2 - G/2 - a_1 Y_1^2 - cL_1. \end{aligned}$$

Assuming that the second-order condition holds, the first-order condition determining the optimal Y_1 is

$$\partial U_1/\partial Y_1 = 1/2 - 2a_1 Y_1^2 = 0$$

which gives the optimal revenue collection

$$Y_1 = 1/4a_1. \qquad (3.8)$$

By symmetry, the optimal revenue collection for region 2 is

$$Y_2 = 1/4a_2. \qquad (3.9)$$

The first derivative of region 1's utility with respect to its lobbying effort is

$$\partial U_1/\partial L_1 = -c < 0. \qquad (3.10)$$

This suggests that the problem has a corner solution of $L_1 = 0$. For future use, we denote this optimal lobbying effort of region 1 defined by equation (3.10) by L_1^H. The superscript H stands for hard budget constraint. Similarly, the optimal lobbying effort of region 2, L_2^H, is also zero.

The equilibrium tax rates can be obtained by substituting equations (3.8) and (3.9) into equations (3.3) and (3.4):

$$x_1 = 1/2 - a_1/2a_2 + 2a_1G \tag{3.11}$$

and

$$x_2 = 1/2 - a_2/2a_1 + a_2G. \tag{3.12}$$

Note that $G_1(0) = G_2(0) = 0$.

These results will be used for comparison with the equilibrium under a soft budget constraint.

REGIONS' LOBBYING UNDER A SOFT BUDGET CONSTRAINT

As mentioned in the introduction, one stylized fact in China's budget formation process is that the regions lobby for additional transfers after the budget plan is officially approved by the Congress, typically in March of each year. In this chapter we label this system as one of soft budget constraint. In such a system, there is no restriction on regions' lobbying for additional transfers after the budget plan is announced by the center. The game under a soft budget constraint involves three stages and proceeds as follows.

(1) In the first stage, the regions determine how much revenue to collect, taking into account that the revenue collection will affect the tax rates. The two regions play a Cournot game in this stage, each taking the other party's action as given when making its own decision.
(2) In the second stage, observing the collected revenues and anticipating the future lobbying of the regions, the center determines the tax rates on regions 1 and 2.
(3) In the third stage, the regions choose their lobbying efforts.

The difference between the game under a hard budget constraint and that under a soft budget constraint (see the previous section) is that the former restricts the regions to choose their lobbying efforts before the center's announcement of the revenue-sharing methods, while in the latter game the regions' total lobbying efforts are determined after the center's announcement of the revenue-sharing methods.

We use backward induction to solve for the equilibrium of this game. First consider region 1's problem in stage 3. Region 1 maximizes its utility by choosing its lobbying effort, L_1:

$$U_1 = (1 - x_1)Y_1 - a_1 Y_1^2 + G_1(L_1) - cL_1. \qquad (3.13)$$

In this problem, Y_1 and x_1 are taken as given, since they are determined in the first and the second stages, respectively. The first-order condition determining the optimal lobbying effort for region 1, L_1^S, is

$$\partial U_1/\partial L_1 = \partial G_1/\partial L_1 - c = 0. \qquad (3.14)$$

The superscript S stands for soft budget constraint. Similarly, the equation defining the optimal lobbying effort for region 2, L_2^S, is

$$\partial U_2/\partial L_2 = \partial G_2/\partial L_2 - c = 0. \qquad (3.15)$$

Comparing equation (3.10) with equation (3.14), and using the assumptions that $G_i(0) = 0$ and $G_i' > 0$, it is easy to see that the optimal lobbying effort under a soft budget constraint is higher than that under a hard budget constraint for both regions:

$$L_1^S > L_1^H \qquad (3.16)$$

and

$$L_2^S > L_2^H. \qquad (3.17)$$

With higher lobbying efforts, the total transfers from the center to the regions will be higher since $G_i(0) = 0$ and $G_i' > 0$. We therefore have the following remark.

Remark: the total expenditure under a soft budget constraint is higher than that under a hard budget constraint.

Next let us look at the center's problem in the second stage. The center chooses x_1 and x_2 to maximize the following objective:

$$U_c = \log[(1 - x_1)Y_1 + G_1(L_1^S)] + \log[(1 - x_2)Y_2 + G_2(L_2^S)] \qquad (3.18)$$

subject to

$$x_1 Y_1 + x_2 Y_2 \geq G + G_1(L_1^S) + G_2(L_2^S)$$

where L_1^S and L_2^S are defined by equations (3.14) and (3.15). Using the same procedure as before, we can solve for the optimal tax rates

$$x_1 = 1/2 - Y_2/2Y_1 + [2G_1(L_1^S) + G]/Y_1 \qquad (3.19)$$

and

$$x_2 = 1/2 - Y_1/2Y_2 + [2G_2(L_2^S) + G]/Y_2. \tag{3.20}$$

With equations (3.19) and (3.20), we can solve for the optimal revenue collections of the two regions in the first stage. Region 1's problem is to choose an optimal Y_1 to maximize its objective:

$$U_1 = (1 - x_1)Y_1 - a_1 Y_1^2 + G_1(L_1^S) - cL_1^S \tag{3.21}$$

where L_1^S is determined by equation (3.14), and x_1 is determined by equation (3.19).

It is easy to see that the optimal revenue collections for the two regions are the same as those under a soft budget constraint:

$$Y_1 = 1/4a_1 \tag{3.22}$$

and

$$Y_2 = 1/4a_2. \tag{3.23}$$

Substituting equations (3.22) and (3.23) into equations (3.19) and (3.20) yields the equilibrium tax rates

$$x_1 = 1/2 - a_1/2a_2 + 4a_1[G_1^S + G/2] \tag{3.24}$$

and

$$x_2 = 1/2 - a_2/2a_1 + 4a_2[G_2^S + G/2] \tag{3.25}$$

where G_1^S and G_2^S are defined by equations (3.14) and (3.15).

Comparing equations (3.24) and (3.25) with equations (3.11) and (3.12), and using inequalities (3.16) and (3.17), we have the following conclusion:

Remark: the tax rates applied to the two regions under a soft budget constraint are higher than those under a hard budget constraint.

In sum, the comparison between the system of a soft budget constraint and the system of a hard budget constraint suggests that a soft budget constraint leads to a higher total expenditure. The intuition is as follows. With a hard budget constraint, the regions have to determine the level of lobbying effort before the tax rates are determined. Because of this timing arrangement, the regions take into account the negative impact of a high lobbying effort on their retained revenue (that is, a higher lobbying effort leads to a higher tax rate). In other words, the tax rates are elastic to the regions' lobbying efforts. By taking into account this negative impact, the regions tend to restrain themselves from lobbying. In the case of a soft budget constraint, however, the regions face predetermined tax rates when they choose their lobbying efforts. An increase in one's lobbying effort will not change the tax rates (that is, the tax rates are inelastic to lobbying efforts in this system). This feature of the soft budget constraint system induces a higher level of lobbying effort compared to that under the system of a hard budget constraint. While the two regions' revenue collections are the same in the two systems, a higher expenditure due to a higher lobbying effort requires higher tax rates on the two regions.

ARE REGIONS BETTER OFF WITH A SOFT BUDGET CONSTRAINT?

From the above analyses, we see that with a soft budget constraint, the regions receive more transfers from the center, but they are penalized by higher tax rates. The question is thus, overall, are the regions better off with a soft budget constraint, compared to the case of a hard budget constraint?

To see the welfare level of region 1 under a hard budget constraint, we simply insert equations (3.8) and (3.11) into equation (3.5):

$$U_1^{\mathrm{H}} = 1/16a_1 + 1/8a_2 - G/2 - cL_1^{\mathrm{H}}.$$

Similarly, the welfare level of region 1 under a soft budget constraint can be obtained by inserting equations (3.22) and (3.24) into equation (3.5):

$$U_1^{\mathrm{S}} = 1/16a_1 + 1/8a_2 - G/2 - cL_1^{\mathrm{S}}.$$

Since $L_1^{\mathrm{S}} > L_1^{\mathrm{H}}$, we have

$$U_1^{\mathrm{H}} > U_1^{\mathrm{S}}.$$

By symmetry, we also have

$$U_2^{\mathrm{H}} > U_2^{\mathrm{S}}$$

where.

$$U_2^{\mathrm{H}} = 1/16a_2 + 1/8a_1 - G/2 - cL_2^{\mathrm{H}}$$

and

$$U_2^{\mathrm{S}} = 1/16a_2 + 1/8a_1 - G/2 - cL_2^{\mathrm{S}}.$$

Remark: the welfare levels of the regions are lower under a soft budget constraint than under a hard budget constraint.

The intuition is as follows. No matter what the total lobbying efforts are (high under a soft budget constraint and low under a hard budget constraint), the regions receive the same amount of retained revenues (including transfers) under the two systems. Region 1's total retained revenue is

$$(1 - x_1)Y_1 + G_1 = 1/2 + 1/8a_2$$

and region 2's total retained revenue is

$$(1 - x_2)Y_2 + G_2 = 1/2 + 1/8a_1.$$

While under a soft budget constraint, the higher lobbying efforts mean higher costs of lobbying. Since additional lobbying costs do not generate additional welfare for the regions, such lobbying is a waste of resources.

From the center's point of view, the two systems yield the same utility:

$$U_C = 2\log(1/8a_1 + 1/8a_2 - G/2).$$

Considering both the center's and the regions' utilities, the system of a soft budget constraint performs strictly worse than the system of a hard budget constraint.

WHY DO THE REGIONS CHOOSE TO LOBBY *EX POST*?

Suppose now that the regions have the choice to lobby only *ex ante* (lobby before the tax rates are set) or to lobby *ex post*. If the regions lobby only *ex ante*, their utility levels are higher compared to the case where they lobby *ex post*. The question we want to ask here is: knowing that the soft budget constraint yields a lower utility level, why in reality do regions choose to lobby *ex post*? In other words, why do they not restrain themselves from lobbying *ex post*?

We claim that the source of this problem is that the regions' lobbying behavior is time inconsistent. *Ex ante*, the regions find it optimal to announce that they will not lobby *ex post*, since the equilibrium with a hard budget constraint gives them higher utility levels. However, if the center has mistakenly believed the regions' promise, and set tax rates according to the calculations based on low lobbying efforts (no *ex post* lobbying), the regions will find it optimal to increase their lobbying efforts *ex post*. The regions' promise of no *ex post* lobbying is therefore not credible. Without an externally imposed mechanism that can restrict *ex post* lobbying the game has only one equilibrium, namely the one of a soft budget constraint.

Formally, the regions' incentive to lobby *ex post* can be shown by looking at the regions' problem after the center has announced the optimal tax rates calculated based on a hard budget constraint. Suppose the regions choose L_1^H and L_2^H in the first stage, and the center mistakenly believes that the regions will not increase their lobbying efforts *ex post*. Let x_1^H and x_2^H denote the two tax rates calculated based on this belief. Region 1's problem in the third stage (after the tax rates are set) is to choose an optimal lobbying effort, L_1, to maximize the objective:

$$U_1 = (1 - x_1)Y_1 - a_1Y_1^2 + G_1(L_1) - cL_1$$

subject to

$$\text{(i) } x_1 = 1/2 - Y_2/2Y_1 + G_1(L_1^H)Y_1 + G/2Y_1$$
$$\text{(ii) } Y_1 = 1/4a_1.$$

The first constraint comes from the optimal reaction of the center in the second stage, and the second constraint is the region's optimal choice of revenue collection in the first stage. Since Y_2 is taken as given (the Cournot assumption), the first-order condition determining the optimal lobbying in the third stage is the same as equation (3.14):

$$\partial U_1/\partial L_1 = \partial G_1/\partial L_1 - c = 0.$$

This equation states that the optimal lobbying effort is $L_1{}^S$, which is higher than the *ex ante* lobbying effort $L_1{}^H (= 0)$. Similarly, region 2 will find it optimal to choose an *ex post* lobbying effort ($L_2{}^S$) higher than the *ex ante* optimal lobbying effort $L_2{}^H (= 0)$.

Remark: *ex post*, the regions find it optimal to increase their lobbying efforts from $L_i{}^H$ to $L_i{}^S$, where $i = 1, 2$. The regions' commitment to a hard budget constraint is not credible. Alternatively, we say that the regions' optimal level of lobbying is time inconsistent.

So far, we have identified that the regions' time-inconsistency problem is the source of too high an expenditure under a soft budget constraint. Before the center's budget plan is determined, the regions find it optimal to commit to a relatively low level of lobbying, since the burden of too much lobbying will be borne by the regions: the tax rates would be high if they lobby intensely. However, after the center has decided the revenue-sharing formulae with the regions, the regions will find it optimal to increase lobbying efforts as the tax rates are no longer elastic to their lobbying efforts. Therefore, the regions' *ex ante* commitment to a low level of lobbying is not credible in the absence of an externally imposed device that prohibits *ex post* lobbying (this system is referred to as one of soft budget constraint). Knowing that the regions are unable to commit and, consequently, the total expenditure will be high, the center sets high tax rates that are compatible with a high level of lobbying in the first place. The equilibrium under a soft budget constraint is therefore a higher level of expenditure, higher tax rates, and a lower welfare level, compared to the case where the regions can commit not to lobby *ex post* (hard budget constraint).

THE COURNOT GAME BETWEEN REGIONS AND LOBBYING EFFECTIVENESS

The above discussion assumed a simple transfer function for each region: transfer from the center to region 1, G_1, depends only on region 1's lobbying effort, L_1, and transfer from the center to region 2, G_2, depends only on region 2's lobbying effort, L_2. This assumption simplifies the analysis since each region's optimal lobbying effort can be independently determined, as in equations (3.14) and (3.15). While this simplification does not affect the qualitative result of regions' time-inconsistent behavior, it is still worthwhile to investigate what can be generated if we adopt a more general transfer function, as used by Tullock (1980) and Becker (1983). This section will derive the Cournot equilibrium using a general specification of transfer function, and show the impact of lobbying effectiveness on the equilibrium outcome.

We now assume that region 1's transfer function depends on both its own lobbying effort, L_1 and region 2's lobbying effort, L_2. Both L_1 and L_2 are greater than or equal to zero. This function has the following properties:

$$(A1)\ G_1(0, L_2) = 0, \qquad (3.26)$$

that is, for any L_2, zero own lobbying effort yields zero transfer;

$$(A2)\ \partial G_1(L_1, L_2)/\partial L_1 > 0 \qquad (3.27)$$

that is, for any L_2, transfer is an increasing function of own lobbying effort;

$$(A3)\ \partial^2 G_1(L_1, L_2)/\partial L_1^2 < 0, \qquad (3.28)$$

that is, the marginal 'productivity' or the effectiveness of lobbying is a decreasing function of own lobbying effort; and

$$(A4)\ \partial^2 G_1(L_1, L_2)/\partial L_1 \partial L_2 < 0, \qquad (3.29)$$

that is, the marginal 'productivity' or the effectiveness of region 1's lobbying is decreasing in region 2's lobbying effort. An assumption similar to this is used in Becker (1983).

The same properties apply to $G_2(L_2, L_1)$, the transfer function of region 2.

We will apply the new transfer function to only the case of a soft budget constraint, for in the case of a hard budget constraint the new specification does not make any difference. The structure of the game under a soft budget constraint is the same as described in the second section. The center's problem and its optimal choice of tax rates are also the same as before. The difference caused by the new transfer function is that the first-order conditions determining regions' optimal lobbying efforts, that is equations (3.14) and (3.15), now become

$$\partial U_1/\partial L_1 + \partial G_1(L_1, L_2)/\partial L_1 - c = 0 \tag{3.30}$$

and

$$\partial U_2/\partial L_2 + \partial G_2(L_2, L_1)/\partial L_2 - c = 0. \tag{3.31}$$

As widely used in the interest group literature (Becker, 1983; Findlay and Wellisz, 1984; van Winden and Frans, 1988), we make the Cournot assumption that each region takes the other region's lobbying effort as given when determining its own lobbying effort. Therefore, the optimal L_1, and L_2 are jointly determined by equations (3.30) and (3.31), the two regions' reaction functions.

Figure 3.1 shows these two reaction functions. Equation (3.30), labeled R_1, is region 1's reaction function to region 2's lobbying effort, L_2, and equation (3.31), labeled R_2, is region 2's reaction to region 1's lobbying effort, L_1. Both reaction functions are downward sloping. To see this, we differentiate equation (3.30):

$$\mathrm{d}L_1/\mathrm{d}L_2 = -[\partial G_1^2(L_1, L_2)/\partial L_1 \partial L_2]/[\partial G_1^2(L_1, L_2)/(\partial L_1)^2] < 0 \tag{3.32}$$

due to assumptions (A3) and (A4). Similarly, for reaction function R_2, we have

$$\mathrm{d}L_2/\mathrm{d}L_1 = -[\partial G_2^2(L_2, L_1)/\partial L_2 \partial L_1]/[\partial G_2^2(L_2, L_1)/(\partial L_2)^2] < 0. \tag{3.33}$$

The stability condition for the Cournot equilibrium is that R_1 has a steeper slope than R_2, that is

$$\begin{aligned}&[\partial G_1^2(L_1, L_2)/(\partial L_1)^2]/[\partial G_1^2(L_1, L_2)/\partial L_1 \partial L_2] > \\ &[\partial G_2^2(L_2, L_1)/\partial L_2 \partial L_1]/[\partial G_2^2(L_2, L_1)/(\partial L_2)^2].\end{aligned} \tag{3.34}$$

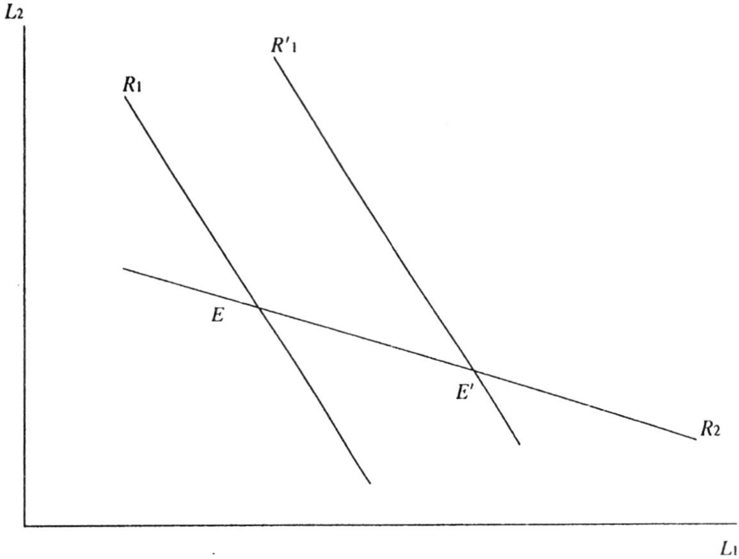

FIGURE 3.1 *Reaction functions of two regions.*

In Figure 3.1, the two reaction functions are depicted as straight lines, only for simplicity. Depending on the property of the transfer function's first and second derivatives, the reaction curves can be either concave or convex. The equilibrium of the two regions' lobbying efforts is given by the point where R_1 cuts R_2.

We now turn to look at the impact of the effectiveness of lobbying on the equilibrium outcome. We have defined that region 1's lobbying effectiveness is its marginal productivity of lobbying, $\partial G_1(L_1, L_2)/\partial L_1$ and region 2's lobbying effectiveness is $\partial G_2(L_2, L_1)/\partial L_2$. In reality, the effectiveness of regions' lobbying depends on various factors: the personal connections between the regional leaders and the leaders in the center, the region's bargaining power (for example, to what extent the region supplies essential goods and services to the center), political factors (for example, the possibility of social unrest due to inadequate transfers), and pressures from local interest groups (for example, large SOEs demanding subsidies) that the region can use. To see the impact of a change in lobbying effectiveness, let us also define that an increase in lobbying effectiveness means a higher marginal productivity of lobbying given the other region's lobbying effort.

Suppose that the two regions' lobbying functions originally have the same effectiveness, so the two reaction functions are symmetric and in equilibrium

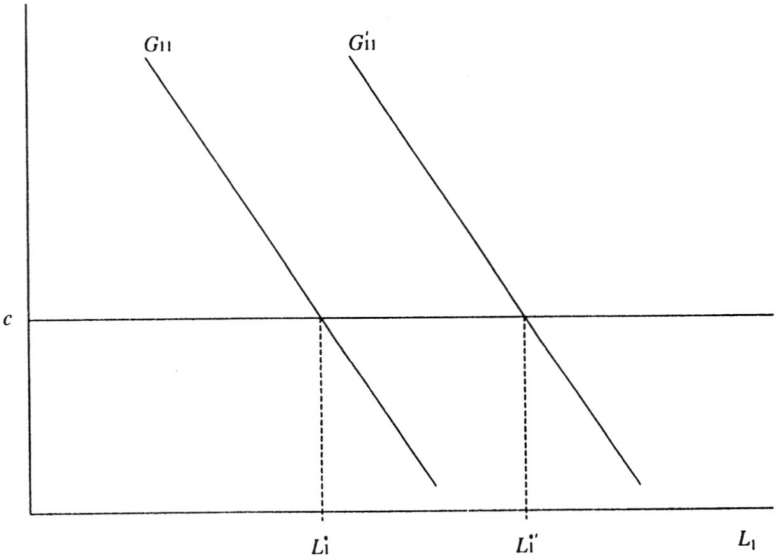

FIGURE 3.2 *The impact of an increase in lobbying effectiveness on optimal L_1.*
NOTE: $G_{11} = \partial G_1(L_1;\ L_2)/\partial L_1;\ G_{11}' = \partial G_1(L_1';\ L_2)/\partial L_1$.

$L_1 = L_2$. An increase in region 1's effectiveness of lobbying increases the equilibrium L_1 for any level of L_2. To see this, let us look at Figure 3.2, which depicts how the optimal L_1 is determined by equation (3.30). The horizontal axis is L_1 and the downward sloping curve is $\partial G_1(L_1, L_2)/\partial L_1$, where L_2 is given. The optimal L_1 is given by L_1^*, which can be read off from the point where the $\partial G_1/\partial L_1$ curve intersects the horizontal line of c, c being the unit cost of lobbying. An increase in region 1's lobbying effectiveness, according to the definition, shifts the $\partial G_1/\partial L_1$ curve upwards. As a result, the optimal level of L_1 increases from L_1^* to $L_1^{*\prime}$. This suggests that an increase in region 1's lobbying effectiveness increases the optimal L_1 for any given level of L_2 (that is, it shifts the reaction function R_1 to the right in Figure 3.1, as labeled by R_1'). Compared to the old equilibrium, E, in Figure 3.1, the new equilibrium, E', has a higher level of L_1 and a lower level of L_2.

Remark: a region with higher effectiveness of lobbying tends to make a higher level of lobbying effort.

It is obvious that if both regions have very low effectiveness of lobbying, the two reaction functions will be located very close to the origin and the equilibrium outcome will involve little lobbying activity. This suggests that an important way to address the problem of regions' excessive lobbying is to reduce their effectiveness by transforming the budget formation procedure from a policy based one to a rule-based one. In other words, the central government's discretion in granting transfers to regions should be restricted by introducing transfer formulae that are based on objective variables and by establishing procedures for budget revision. For example, the Congress should regulate that any large amount of transfers exceeding the originally approved expenditure plan must be rectified by a special committee of the Congress. Political reform in the direction of increasing the publicity of government transfers will also constitute an important impediment to the regions' excessive lobbying. Privatizing SOEs (to increase the independence of these firms from government protection) is also a way to reduce local pressures for transfers.[5]

THE BUDGET CONSTRAINT PROBLEM UNDER THE CENTER'S COMMITMENT

In previous sections, we assume that the center does not commit to a fixed revenue-sharing method (that is, the center determines the tax rates after observing the regions' revenue collection). As discussed in Chapter 2, this assumption describes the characteristics of the current Chinese fiscal system. Based on this assumption, we have explicitly shown that the outcome under a hard budget constraint is strictly better than that under a soft budget constraint. Compared to the system of soft budget constraint, the system of hard budget constraint yields a lower level of expenditure, lower tax rates, and avoids the resource waste associated with excessive lobbying. However, the outcome of a soft budget constraint is the only subgame perfect equilibrium in the game since the regions face a time-inconsistency problem.

This section asks what would happen to regions' lobbying activities if the central–local revenue-sharing system is reformed so that the center does commit to its pre-announced tax rates (that is, the center determines the tax rates before observing the regions' revenue collection). It is shown that the regions' time-inconsistency problem remains even in the case of the center's pre-commitment to tax rates. The following two subsections discuss the case of a soft budget constraint under the center's commitment and the case of a hard budget constraint under the center's commitment, respectively.

Commitment with a soft budget constraint

In a commitment regime, the center determines the tax rates before the regions collect the revenues. Under a soft budget constraint, the regions choose their lobbying efforts after the center's announcement of the tax rates. Therefore, in the case of commitment with a soft budget constraint, the center moves first and the regions choose their two instruments second. The game thus has two stages and proceeds as follows.

(1) In the first stage, the center determines the tax rates on regions 1 and 2.
(2) In the second stage, the regions determine how much revenue to collect and choose their lobbying efforts.

Region 1's problem in the second stage is to choose Y_1 and L_1 to maximize the objective:

$$U_1 = (1 - x_1)Y_1 - a_1 Y_1^2 + G_1(L_1) - cL_1 \tag{3.35}$$

where x_1 and L_2 are taken as given. $G_1(L_1)$ is the transfer function defined in the first section. The optimal revenue for region 1 is

$$Y_1 = (1 - x_1)/2a_1. \tag{3.36}$$

Similarly, the optimal revenue for region 2 is

$$Y_2 = (1 - x_2)/2a_2. \tag{3.37}$$

The optimal lobbying effort, L_1^S, is defined by the first-order condition (the same as equation (3.14)):

$$\partial U_1/\partial L_1 = \partial G_1/\partial L_1 - c = 0. \tag{3.38}$$

The superscript S stands for soft budget constraint. Similarly, the equation defining the optimal lobbying effort for region 2, L_2^S, is

$$\partial U_2/\partial L_2 = \partial G_2/\partial L_2 - c = 0. \tag{3.39}$$

In the first stage, the center takes into account equations (3.36)–(3.39) and chooses x_1 and x_2 to maximize the objective

$$U_c = \log[(1 - x_1)Y_1 + G_1(L_1^S)] + \log[(1 - x_2)Y_2 + G_2(L_2^S)] \tag{3.40}$$

subject to

$$\text{(i) } x_1 Y_1 + x_2 Y_2 \geq G + G_1(L_1^S) + G_2(L_2^S),$$
$$\text{(ii) } Y_1 = (1 - x_1)/2a_1,$$
$$\text{(iii) } Y_2 = (1 - x_2)/2a_2$$

where the right-hand side of constraint (i) is a constant, since L_1^S and L_2^S are determined by equations (3.38) and (3.39). Constraints (ii) and (iii) are the regions' reaction functions given by equations (3.36) and (3.37).

The detailed procedure for solving this problem is given in Appendix 3.1. Since the explicit solution to this problem is difficult to obtain, we simply denote the two optimal tax rates under commitment by

$$x_1 = x_1^c[G + G_1(L_1^S) + G_2(L_2^S)] \tag{3.41}$$

and

$$x_2 = x_2^c[G + G_1(L_1^S) + G_2(L_2^S)] \tag{3.42}$$

where the superscript c on x_i stands for commitment.

Compared to the case of commitment with a hard budget constraint, which will be discussed in the next subsection, the case of commitment with a soft budget constraint yields a higher level of expenditure and a lower level of revenue collection.

Commitment with a hard budget constraint

This case differs from the case of commitment with a soft budget constraint in that the regions are restricted to choosing their lobbying efforts before the center sets the tax rates. Formally, the game has three stages and proceeds as follows.

(1) In the first stage, the regions choose their lobbying efforts.
(2) In the second stage, the center determines the tax rates on regions 1 and 2, given the total expenditure calculated from regions' lobbying efforts in the first stage.
(3) In the third stage, the regions determine how much revenue to collect, given the tax rates set in the second stage.

Region 1's problem in the third stage is to choose an optimal Y_1 to maximize the objective

$$U_1 = (1 - x_1)Y_1 - a_1 Y_1^2 + G_1(L_1^H) - cL_1{}^H \tag{3.43}$$

where x_1 is determined in the second stage and L_1^H and L_2^H are determined in the first stage. Region 1's optimal revenue is

$$Y_1 = (1 - x_1)2a_1. \tag{3.44}$$

Similarly, the optimal revenue for region 2 is

$$Y_2 = (1 - x_2)2a_2. \tag{3.45}$$

In the second stage, the center choose x_1 and x_2 to maximize the objective

$$U_c = \log[(1 - x_1)Y_1 + G_1(L_1^H)] + \log[(1 - x_2)Y_2 + G_2(L_2^H)] \qquad (3.46)$$

subject to

$$\text{(i) } x_1 Y_1 + X_2 Y_2 \geq G + G_1(L_1^H) + G_2(L_2{}^H),$$
$$\text{(ii) } Y_1 = (1 - x_1)2a_1,$$
$$\text{(iii) } Y_2 = (1 - x_2)2a_2$$

where the right-hand side of constraint (i) is a constant, since L_1^H and L_2^H are determined in the first stage. Constraints (ii) and (iii) are the regions' reaction functions given by equations (3.44) and (3.45).

Again, readers are referred to Appendix 3.1 for how to solve the center's problem. We denote the center's optimal tax rates by

$$x_1 = x_1^c[G + G_1(L_1^H) + G_2(L_2^H)] \qquad (3.47)$$

and

$$x_2 = x_2^c[G + G_1(L_1^H) + G_2(L_2^H)] \qquad (3.48)$$

where the superscript c on x_i stands for commitment. One can see that the functional form of equations (3.47) and (3.48) are the same as equations (3.41) and (3.42). The difference between the optimal tax rates in this subsection and the previous subsection is that the lobbying efforts differ.

Now let us look at region 1's problem in the first stage. Region 1 chooses L_1^H to maximize the objective

$$U_1 = (1 - x_1)Y_1 - a_1 Y_1^2 + G_1(L_1^H) - cL_1^H \qquad (3.49)$$

subject to

$$\text{(i) } x_1 = x_1^c[G + G_1(L_1^H) + G_2(L_2^H)],$$
$$\text{(ii) } Y_1 = (1 - x_1)/2a_1.$$

One can see that constraint (i) is given by equation (3.47) derived from the center's problem in the second stage. Constraint (ii) is equation (3.44) derived from the region's problem in the third stage.

To characterize the optimal lobbying effort, we need to first look at the relationship between the total expenditure and the tax rates. From Appendix 3.1 we see that a higher transfer yields a higher own tax rate:

$$dx_1/dG_1 > 0. \qquad (3.50)$$

From assumption (A2) in the first section, we also know that a higher lobbying effort yields a higher transfer:

$$\partial G_1/\partial L_1^{\mathrm{H}} > 0. \tag{3.51}$$

Combining the above two results, we have

$$\partial x_1/\partial L_1^{\mathrm{H}} > 0,$$

that is, a higher lobbying effort yields a higher own tax rate.

The first-order condition determining region 1's optimal lobbying effort is

$$\partial U_1/\partial L_1^{\mathrm{H}} = -(1 - x_1)(\partial x_1/\partial L_1^{\mathrm{H}})/2a_1 + \partial G_1/\partial L_1^{\mathrm{H}} - c = 0. \tag{3.52}$$

Note that the first term in equation (3.52) is negative, for $x_1/L_1^{\mathrm{H}} > 0$. Compared to equation (3.38) in the case of commitment with a soft budget constraint, one can see that the case of commitment with a hard budget constraint yields a lower lobbying effort because

$$\partial U_1/\partial L_1^{\mathrm{H}} < \partial U_1/\partial L_1^{\mathrm{S}}.$$

Similarly, the first-order condition for region 2's optimal lobbying effort is

$$\partial U_2/\partial L_2^{\mathrm{H}} = -(1 - x_2)(\partial x_2/\partial L_2^{\mathrm{H}})/2a_2 + \partial G_2/\partial L_2^{\mathrm{H}} - c = 0. \tag{3.53}$$

Compared to equation (3.39), one sees that the optimal lobbying effort of region 2 is also lower in the case of commitment with a hard budget constraint than in the case of commitment with a soft budget constraint.

Remark: as a result of the lower lobbying efforts, the total expenditure in equilibrium is lower in the case of commitment with a hard budget constraint than in the case of commitment with a soft budget constraint.

As a result of lower lobbying efforts, total expenditure is lower [see equation (3.51)], and tax rates are lower [see equation (3.50)]. From equations (3.44) and (3.45), we have the following remark.

Remark: the regions collect more revenues in the case of commitment with a hard budget constraint than in the case of commitment with a soft budget constraint.

From the timing assumption of the game, we also have the following remark.

Remark: as a Stackelberg leader in the game under a hard budget constraint (the regions choose lobbying efforts before the center sets the tax rates), the utility levels of the regions are higher than in the case of a soft budget constraint.

The question is then, knowing that sticking to a low level of lobbying effort (that is, committing not to lobby *ex post*) yields a higher utility level, why do the regions choose to lobby *ex post*? The answer to the same question in the fourth section applies here. *Ex ante*, the regions do intend to commit to *ex ante* lobbying, since too much lobbying will increase their tax rates. However, after the center has set the tax rates, the regions will have the incentive to increase their lobbying effort because the tax rates are no longer elastic to the lobbying efforts. The regions face the same time-inconsistency problem as discussed in the fourth section. Without a mechanism that can enforce the regions' commitment to *ex ante* lobbying, the only equilibrium of this game is one of a soft budget constraint.

CONCLUDING REMARKS

This chapter analyzes the process of expenditure target formation and identifies the source of the excessive expenditure expansion in the framework of a central–local game. It is shown that under the system of a soft budget constraint – a system that does not prohibit regions' lobbying for transfers from the center after the budget is officially approved by the Congress – the regions make too much lobbying efforts and result in excessive expenditure expansion, compared to the system of a hard budget constraint in which such *ex post* lobbying is prohibited. The source of the excessive expenditure expansion under a soft budget constraint is that the regions' behavior is time inconsistent. This result holds whether the center commits to the pre-announced fixed revenue-sharing method or not.

The analysis of the model suggests that an important way to address the problem of regions' excessive lobbying is to reduce the lobbying effectiveness by transforming the budget formation procedure from a policy - based one to a rule-based one. In other words, the central government's discretion in granting transfers to regions should be strictly restricted and the budget formation and revision should be subject to well-formulated criteria and procedures. For example, the Congress should regulate that any large amount of transfers exceeding the originally approved expenditure plan must be ratified by a special committee of the Congress. Political reform in the direction of increasing the publicity of government transfers will also constitute an important impediment to the regions' excessive lobbying. Privatizing SOEs (to increase the independence of these firms from government protection) will also help reduce local pressures for transfers.

As in many studies on rent-seeking behavior, implicitly assumed in this chapter is that the center has incomplete information about regions' actual expenditure needs so that it has to rely on regions' lobbying to determine transfers. This assumption itself suggests another implication: establishing more accurate measures of regions' expenditure needs and the procedures to apply such measures will help reduce the degree of information incompleteness. With a better informed center, intergovernmental transfers from the center to the regions are more likely to be determined based on criteria other than intensity of lobbying, and the regions' incentive to lobby will decrease accordingly.

APPENDIX 3.1: SOLVING FOR OPTIMAL COMMITMENT TAX RATES FOR A GIVEN EXPENDITURE TARGET

This model assumes that the center chooses the tax rates in the first stage and regions determine their revenue collection in the second stage (after observing the tax rates). The center's problem in the first stage is to choose x_1 and x_2 to maximize the objective

$$U_c = \log\left[(1-x_1)Y_1 + G_1\right] + \log\left[(1-x_2)Y_2 + G_2\right] \tag{A3.1}$$

subject to

$$x_1 Y_1 + x_2 Y_2 \geq G + G_1 + G_2 \tag{A3.2}$$

where G is the expenditure by the center itself, G_1 is the transfer to region 1 and G_2 is the transfer to region 2. For the purpose of analyzing the effect of a change in G_i on tax rates, we also assume that G_1 and G_2 are exogenously given. Also assume that $x_i \leq 1,\ i = 1, 2$.

We know from the regions' problem that, given x_1 and x_2 fixed over time (that is, commitment), the regions choose the following levels of revenue collection:

$$Y_1^c = (1-x_1)/2a_1 \tag{A3.3}$$

and

$$Y_2^c = (1-x_2)/2a_2. \tag{A3.4}$$

Substituting equations (A3.3) and (A3.4) into equation (A3.1) and specifying the Lagrange

$$\begin{aligned} L = {} & \log\left[(1-x_1)(1-x_1)/2a_1 + G_1\right] + \log\left[(1-x_2)(1-x_2)/2a_2 + G_2\right] \\ & + \lambda\left[x_1(1-x_1)2a_1 + x_2(1-x_2)/2a_2 - G - G_1 - G_2\right] \end{aligned}$$

we obtain the first-order conditions

$$\begin{aligned}\partial L/\partial x_1 = -(1-x_1)/\{a_1[(1-x_1)^2/2a_1 + G_1]\} \\ + \lambda(1-2x_1)/2a_1 = 0\end{aligned} \tag{A3.5}$$

and

$$\begin{aligned}\partial L/\partial x_2 = -(1-x_2)/\{a_2[(1-x_2)^2/2a_2 + G_2]\} \\ + \lambda(1-2x_2)/2a_2 = 0.\end{aligned} \tag{A3.6}$$

Eliminating λ from the above conditions gives

$$\begin{aligned}[(1-x_1)^2/2a_1 + G_1](1-2x_1)/(1-x_1) = \\ [(1-x_2)^2/2a_2 + G_2](1-2x_2)/(1-x_2).\end{aligned} \tag{A3.7}$$

Note that constraint (A3.2) implies

$$G + G_1 + G_2 = (1-x_1)x_1/2a_1 + (1-x_2)x_2/2a_2. \tag{A3.8}$$

Equations (A3.7) and (A3.8) define the optimal tax rates. However, solving these two equations yields a third-order polynomial equation. It is easy to see that both x_1 and x_2 must be less than 1/2, otherwise the left-hand sides of equations (A3.5) and (A3.6) will be negative.

To see the impact of a change in G_1 on the optimal tax rates, we totally differentiate equations (A3.7) and (A3.8). Rearranging terms in equation (A3.7) gives

$$\begin{aligned}(1-3x_1+2x_1^2)/2a_1 + G_1(1-2x_1)/(1-x_1) = \\ (1-3x_2+2x_2^2)/2a_2 + G_2(1-2x_2)/(1-x_2).\end{aligned}$$

The differential equation corresponding to equation (A3.7) is

$$\begin{aligned}(-3+4x_1)\mathrm{d}x_1/2a_1 + \mathrm{d}G_1(1-2x_1)/(1-x_1) - G_1\mathrm{d}x_1/(1-x_1)^2 = \\ (-3+4x_2)\mathrm{d}x_2/2a_2 + \mathrm{d}G_2(1-2x_2)/(1-x_2) - G_2\mathrm{d}x_2/(1-x_2)^2\end{aligned}$$

that is,

$$\begin{aligned}\mathrm{d}x_1[(-3+4x_1)/2a_1 - G_1\mathrm{d}x_1/(1-x_1)^2] + \mathrm{d}G_1(1-2x_1)/2(1-x_1) = \\ \mathrm{d}x_2[(-3+4x_2)/2a_2 - G_2\mathrm{d}x_2/(1-x_2)^2] + \mathrm{d}G_2(1-2x_2)/2(1-x_2).\end{aligned} \tag{A3.9}$$

The differential equation corresponding to equation (A3.8) is

$$\mathrm{d}G_1 + \mathrm{d}G_2 = (1-2x_1)\mathrm{d}x_1/2a_1 + (1-2x_2)\mathrm{d}x_2/2a_2. \tag{A3.10}$$

Setting $dG_2 = 0$, we have

$$dx_2 = [dG_1 - (1 - 2x_1)dx_1/a_1]a_2/(1 - 2x_2). \tag{A3.11}$$

Substituting equation (A3.11) into equation (A3.9) we have

$$dx_1[(-3 + 4x_1)/2a_1 - G_1dx_1/(1 - x_1)^2 + dG_1(1 - 2x_1)/(1 - x_1) =$$
$$[dG_1 - (1 - 2x_1)dx_1/a_1]a_2/(1 - 2x_2)[(-3 + 4x_2)/2a_2 - G_2dx_2/(1 - x_2)^2].$$

After rearranging terms we can see that

$$dx_1/dG_1 > 0$$

since

$$(-3 + 4x_1)/2a_1 - G_1dx_1/(1 - x_1)^2 < 0,$$
$$(-3 + 4x_1)/2a_2 - G_2dx_2/(1 - x_2)^2 < 0,$$
$$\frac{a_2(1 - 2x_1)}{a_1(1 - 2x_2)} > 0,$$

and

$$(1 - 2x_1)/(1 - x_1) > 0.$$

Similarly, we have

$$dx_2/dG_2 > 0.$$

APPENDIX 3.2: A COMPARISON WITH TULLOCK (1980) AND BECKER (1983)

This chapter has adopted a transfer function that is similar to those used in Tullock (1980) and Becker (1983) (hereafter T&B). However, the distinct game structure in this chapter allows us to analyze the institutional choice to limit excessive lobbying, while this is not possible in T&B. This appendix briefly reviews the game structures in T&B and indicates the main difference between these models and my model.

In a streamlined Tullock's (1980) model, there are two competitors, *A* and *B*, competing for a payoff, *P*. To compare with our model, the payoff can be interpreted as a transfer from the government. Each competitor maximizes her expected net present value, V_i, where $i = A,B$. *A*'s net present value is given by

$$V_A = aPe^{-\theta t}/(a + b) - a \tag{A3.12}$$

where *a* and *b* are rent-seeking expenditures (in our case, lobbying efforts) by

A and B, P is the payoff at time t, θ is the time discount rate and $a/(a+b)$ is the probability function of winning. Player B has a symmetric value function.

To determine the expenditure by each player a Cournot reaction process is employed, whereby equation (A3.12) is differentiated and the derivative set equal to zero. This results in a Cournot–Nash equilibrium solution whereby each competitor spends

$$a = b = Pe^{-\theta t}/4. \tag{A3.13}$$

In a simplified Becker's (1983) model, there are two interest groups in the society, s and t. s is lobbying for government subsidies. t, which pays the taxes to finance the subsidies, is lobbying against such subsidies. The amount of taxes on t, or subsidies to s, is determined by an influence function that depends on the pressure, p, exerted by s and t and other variables (x):

$$S = T = I(p_s, p_t, x) \tag{A3.14}$$

where S and T are subsidies to s and taxes on t, respectively. The government has to balance its budget by setting $S = T$. The influence function has the following properties:

$$\partial I/\partial p_s > 0 \quad \text{and} \quad \partial I/\partial p_t > 0.$$

That is, increased pressure by s raises subsidies and increased pressure by t lowers subsidies.

Each group's pressure is an increasing function of its resources spent on lobbying, a_t and a_s. That is,

$$p_t = p_t(a_t) \tag{A3.15}$$

and

$$p_s = p_s(a_s) \tag{A3.16}$$

where $p'_t > 0$ and $p'_s > 0$. Each group maximizes its total income, which includes (excludes) the subsidies (taxes). The total income of the two groups are defined by

$$Z_s = Z_s^0 + R_s - a_s \tag{A3.17}$$

and

$$Z_t = Z_t^0 + R_t - a_t \tag{A3.18}$$

where Z_s^0 and Z_t^0 are the initial levels of income of the two groups. R_s is the actual transfer received by the s group, excluding deadweight losses and expenditures to control free riding within the group. R_s is an increasing function of I, that is $R_s = R_s(I)$ with $R'_s > 0$, but $R_s \leq I$. Similarly, R_t is the actual amount of money taken away from group t, including the deadweight losses within group t. It is an increasing function of I, that is $R_t = R_t(I)$ with $R'_t > 0$, but $R_t \geq I$.

The two groups play a Cournot game and determine their optimal spending on lobbying by solving the following equations:

$$dR_s/da_s = (\partial R_s/\partial I)(\partial I/\partial p_s)(\partial p_s/\partial a_s) = 1 \qquad \text{(A3.19)}$$

and

$$dR_t/da_t = -(\partial R_t/\partial I)(\partial I/\partial p_t)(\partial p_t/\partial a_t) = -1. \qquad \text{(A3.20)}$$

Comparative static exercises can show how the properties of R_i, p_i (where $i = s,\ t$) and I influence the equilibrium outcome of lobbying spending of the two groups.

Compared with my model, these two models are very similar to the case of a soft budget constraint, in which two regions (interest groups) lobby for transfers without considering the feedback from the center (government). Implicitly assumed in T&B is that the government has no ability to reallocate resources after the lobbying-based transfers are made. Based on this assumption, T&B's focus is given to the Cournot interaction between the interest groups in a single stage. The government plays no role in this game.

My model extends the game into a multistage one, in which the center is explicitly modeled and the timing of the center's move and the regions' move plays an important role in determining the outcome. While retaining the Cournot game structure between the regions, I allow the center to play a Stackelberg game with the two regions. When the regions are allowed to lobby after the center has determined the tax rates, the center acts as a Stackelberg leader and the regions act as followers. This system is referred to as one of a soft budget constraint. When the regions are restricted to lobby before the center determines the tax rates, the regions act as the Stackelberg leader and the center acts as the follower. This system is referred to as one of a hard budget constraint. I showed that the system of a hard budget constraint yields lower lobbying efforts and higher welfare levels for the regions compared to the system of a soft budget constraint. This result suggests that an important institutional reform to address the problem of excessive lobbying is to establish a mechanism that can force the regions to commit to *ex ante* lobbying. In T&B, however, this option is not open since the interaction between the government and the interest groups is not spelt out. Moreover, with the Stackelberg game structure, I am able to show that the source of the regions' excessive lobbying is the regions' time-inconsistent behavior – this conclusion cannot be derived from a simple one-stage Cournot game.

4 Intergovernmental Relations and Monetary Management

INTRODUCTION

As pointed out in Chapter 1, fiscal policy in China has virtually played no role in stabilizing the economy since the central fiscal authority has faced increasing fiscal imbalance and was overburdened by other objectives. During the course of the decentralization, the main task for macroeconomic control has been performed by the monetary policy. Unfortunately, the monetary policy has not functioned well: it has failed to stabilize the economy when necessary and, on many occasions, it became the cause of macroeconomic instability. Repeated inflationary pressures occurred during the 15-year reform period, with the retail price index shooting up to 18.7 per cent in 1989 and 21.8 in 1994. At the same time, output has also fluctuated significantly and frequently.

This chapter argues that the central bank *per se* is not to be blamed for the excessive monetary growth that created repeated high inflation; rather, the institutional features of the current central–local monetary relations generate this result endogenously. A central–local monetary game is presented as a theory to explain how regions take advantage of the central bank's lack of commitment to a pre-announced credit policy by forcing the central bank to revise the credit ceiling upwards, thus creating inflation. This game also explains how reform cycles – the alternating of decentralization and recentralization – over the past decade could be a consequence of the central–local monetary interaction and why reform cycles have coincided with the monetary cycles.

This chapter will also examine the implications of the 1994 monetary reform. This reform, if successfully implemented, will address a number of important drawbacks of the old system that led to the central bank's weak control over money supply. In particular, the establishment of the new central bank branch system, the centralization of personnel management of the central bank branches and specialized banks, and the separation of policy lending from commercial lending are all in the right direction. Nevertheless, further reforms towards a more independent central bank, the application of indirect measures for monetary control, and the elimination of various structural distortions that led to a conflict of interest between the center and the localities will be necessary for an effective monetary management system in the 1990s.

The first section describes the evolution of China's financial sector development before 1994. The second section provides an overview of the central bank's monetary control mechanism under the decentralized banking system. The third section presents some evidence of macroeconomic fluctuation and its link with the money supply. Using a game theoretic framework, the fourth section discusses how China's decentralized banking system has substantially weakened the central bank's ability to control money supply. The fifth section extends the central–local monetary game to explain reform cycles in China. The sixth and seventh sections look at the 1994 monetary reform and its implications for the central bank's monetary control. The last section concludes.

FINANCIAL SECTOR DEVELOPMENT BEFORE 1994[1]

Along with the decentralization and market-oriented reforms in other aspects of the economy, China's banking system has also changed significantly since the early 1980s. Before 1983, most investments in fixed assets were direct transfers or grants from the government budget. In 1983, direct grants were replaced with interest-bearing loans to agriculture, construction, and production enterprises in an attempt to solve the soft-budget problem of state enterprises. Consequently, the banking system gradually became the primary channel through which investments were financed and the central authorities exercised macroeconomic control. At the same time, the importance of budgetary expenditure in economic adjustment has declined rapidly.

Various specialized banks were created or re-established during the 1980s. The Agriculture Bank, the Construction Bank, and the Bank of China were separated from the operations of People's Bank of China (PBC, or the central bank). Each of these institutions was to provide service to a designated sector in the economy. This approach prevented competition between the banking institutions and required the prospective customers – the peasants, the industrial enterprises, and the trade or foreign-invested companies – to deal with a single institution.

China's banking system now comprises a wide variety of specialized banks. Under the PBC, the major specialized banks and non-bank institutions are the Agricultural Bank, the Bank of China, the Construction Bank, the Industrial and Commercial Bank, the Rural Credit Cooperative, the Urban Credit Cooperative, the Bank of Communications, the China International Trust and Investment Corporation (CITIC), the Guangda Finance Corporation, and the People's Insurance Company.

The main duty of the Agricultural Bank is to manage the budget for agriculture – receiving deposits and extending loans to agricultural production projects. The Construction Bank manages government grants for fixed assets investment and extends loans to fixed assets investment. The Industrial and Commercial Bank receives deposits and provides working capital for enterprises. The Bank of China has primarily focused on deposits and loans of foreign exchange and international transactions.[2] In recent years, the restriction over the business scope of the specialized banks has been somewhat relaxed. The scope of business for the Bank of Communications is not limited to a single line of commerce and the institution offers financial services throughout the world.

Many non-bank financial institutions and local commercial banks also emerged in the 1980s and began to operate and compete with the specialized banks for financial resources. The major non-bank financial institutions include the CITIC, the Guangda Finance Corporation, the People's Insurance Company, and numerous investment companies established by specialized banks and large enterprise groups. Besides these non-bank financial institutions, many local commercial banks located in and serving provinces or the Special Economic Zones (SEZs) have been established since 1987. Examples include the Guangdong Development Bank, Bank of China Merchants (Shekou), Shenzhen Development Bank, and Pudong Development Bank. Many small private banks (Siren Qianzhuang) also operate in China. These banks are the result of entrepreneurs pooling their funds and borrowing additional money to begin operations. The private banks typically pay higher interest rates on deposits than the other major institutions, and they charge higher loan rates than the major banks. Although illegal, these private banks are not officially sanctioned by the government.

China's banking system expanded rapidly during the reform period. The annual aggregate deposits, loans, and assets of the banking system grew at strikingly high rates. During 1979–93, the banking system's loans, deposits, and total assets grew at annual average rates of 20.1, 22.6, and 22.7 per cent, respectively. These rates far exceeded the record of most other developing countries during the same period, even after taking inflation into account. For this period, total deposits increased from Yuan 133.9 billion to Yuan 2 323.0 billion, while total loans increased from Yuan 204.0 billion to Yuan 2 646.1 billion. In 1993, total loans from the banking system were five times government budgetary expenditure, reflecting the rapidly increasing role of banking system in the economy.[3]

While the increasing magnitude of the banking activities in the course of decentralization reform provided a strong stimulus to the growth of the economy, the mechanism of effective control over money supply is far from being well established. The weak monetary control is largely responsible for the repeated economic overheating that characterized the reform period. The next section starts the discussion on this issue with China's monetary control mechanism.

THE CENTRAL BANK'S INSTRUMENTS OF MONETARY CONTROL

Officially, the PBC's role is to formulate and implement monetary policies. However, since the PBC is only a ministry-level body under the State Council, all major monetary policies, including the credit plan and the cash issuance plan, are subject to approval by the State Council. The responsibility of the central bank is therefore to carry out the central government's monetary policy and police all the financial institutions according to the financial discipline of the government.

The relationship between the central bank and specialized banks and their local branches can be described as follows.

(1) All specialized banks and their local branches are subject to the control of the credit plan formulated by the central bank. The credit plan determines the credit ceiling for each specialized bank and each of its local branches; a credit ceiling defines the total loans each specialized bank and each of its local branches can extend within a year. The specialized banks use the funds they control (the deposits they obtained, their own capital, plus the amount of money they are allowed to borrow from the central bank according to the credit plan) to make loans.

(2) The specialized banks submit a certain proportion of their deposits to the central bank as the required reserve. The central bank may change the reserve rate as an instrument of monetary policy. In 1985, the required reserve ratio was 10 per cent. The ratio was raised to 13 per cent in 1988. In 1992 an additional excess reserve ratio of 7 per cent was applied, bringing the total effective reserve ratio to 20 per cent. However, the reserve ratios were not effective because specialized banks have generally held excess reserves. In most cases, the binding constraint that the specialized banks face is the credit plan.

(3) The central bank strictly controls the interest rates for deposits and lending in all specialized banks. The rates are uniform across banks and vary according to types of projects. In accordance with the industrial policy, lending to the agricultural, infrastructure, and energy sectors are

in general subject to low rates. The central bank occasionally adjusts the official interest rates as an instrument for demand control. However, the effectiveness of this instrument is limited since the official interest rates are often negative and the state-owned enterprises (SOEs)' borrowing is often insensitive to interest rates.

(4) The central bank extends credit to specialized banks which fall short of funds for local loans or meeting the reserve requirement. The lending rate for such credit is occasionally used as an instrument of money supply control.

The central bank's most important instrument to control money supply is the formulation and implementation of the credit plan. Although the central bank also formulates a cash issuance plan every year parallel to the credit plan, the cash supply is in effect endogenous as long as the total credit supply is determined. In particular, the demand for cash comes largely from wage payments and agricultural procurement. The former has to do with how much money is lent and the latter is a 'hard' demand, none of which is at the central bank's discretion. We will therefore focus on the credit plan management in the following discussion.

The overall credit plan is formulated by the central bank under the direction of the State Council and in cooperation with the State Planning Commission. The process of making a credit plan can be summarized as 'from the top to the bottom, then from the bottom to the top'. The first phase (from the top to the bottom) is a process of formulating the credit plan starting from the central bank. The money supply is calculated by the central bank according to the forecasted economic growth, price level, and other factors such as the velocity of money and the monetization process. The tentative credit plan includes the total size of new loans and the allocation of the total credit among the specialized banks. The central bank then informs the specialized banks of this tentative plan and asks them for feedback. The headquarters of specialized banks allocate the quotas and required deposit targets to their provincial and regional branches. Then the provincial and regional specialized banks make their own plans.

The next phase is from the bottom to the top. The provincial branches of the central bank summarize the deposit and credit plans of all local specialized bank branches and formulate the provincial plans for deposits, loans, and cash issuance. Then all provincial branches of the central bank submit their plans to the headquarters. Based on these provincial plans, the central bank's headquarters revises its original plan and finalizes the overall credit plan. Once the plan is approved, it becomes the yardstick according to which the monetary policy is conducted.[4]

MONEY SUPPLY AND MACROECONOMIC STABILITY

Between 1979 and 1994, the economy experienced four expansions and four contractions in terms of GNP growth (see Figure 4.1). Variation in production growth has been significant: the peak of GNP growth rates during the period 1979–94 reached 14.5 per cent in 1984 while the trough was approximately 4 per cent in 1990. At the same time, the magnitude of price change has also been large, compared to those in many other countries. In 1983, the increase of the retail price index was only 1.5 per cent. In 1988, this price index went up by 18.7 per cent. The increase of this index declined to 2.1 per cent in 1990 and resurged to 21.7 per cent in 1994 (see Figure 4.2).

The output and price movements are largely created by fluctuation in money supply, rather than exogenous supply or demand shocks. The positive link between money supply and output growth is rather clear, as pointed out by a number of articles (Feltenstein and Ha, 1989; Walker, 1989; Chan and Deaves, 1992; and so on). The price change is also positively correlated to money supply, but to a lesser extent. This is because some of the short-term price increases, such as that at end of 1988, could be attributed to consumers' expectations of a price increase due to the government's announcement of price reform. When consumers' expectation of a price increase is high, they have the incentive to rush to purchase consumer goods, thereby increasing the velocity of money. Nevertheless, some econometric analyses show that

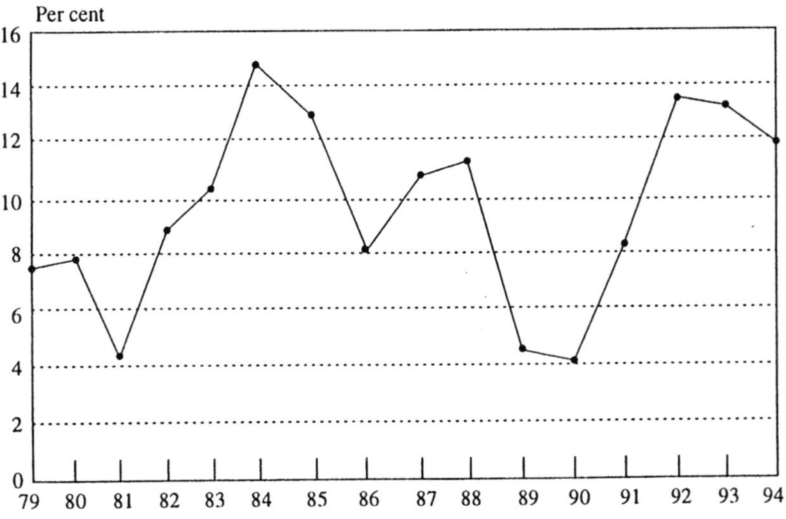

FIGURE 4.1 *Growth rates of real GNP, 1979–94.*

the long-term relationship between price movement and the growth of money supply in China is still rather stable (Chow, 1987).

The central bank's money supply decisions seem not to have stabilized the economy; rather, in most cases, they were causes of the economic cycles. During all periods of economic overheating, the credit and cash issuance plans were not strictly adhered to and were repeatedly revised upwards by substantial margins. When inflation rises to a level where political instability becomes an obvious signal, the central authority turns to reduce the credit and cash issuance sharply. The economy then enters a period of recession. This type of stop-and-go cycle has been a stylized feature of the Chinese monetary control over the past decade and a half, and has been well documented in the literature.[5]

While the links between output, price, and money supply have been relatively well established, the cause of these variables' excessive growth during periods of economic overheating is less clear. There are several popular explanations among Chinese economists and the policy-making circle, however (Liu *et al.*, 1994).

(1) Excessive investment. This is an opinion often heard from government officials and economists: excessive investment leads to excessive demand and creates inflation. However, it is easy to see from the

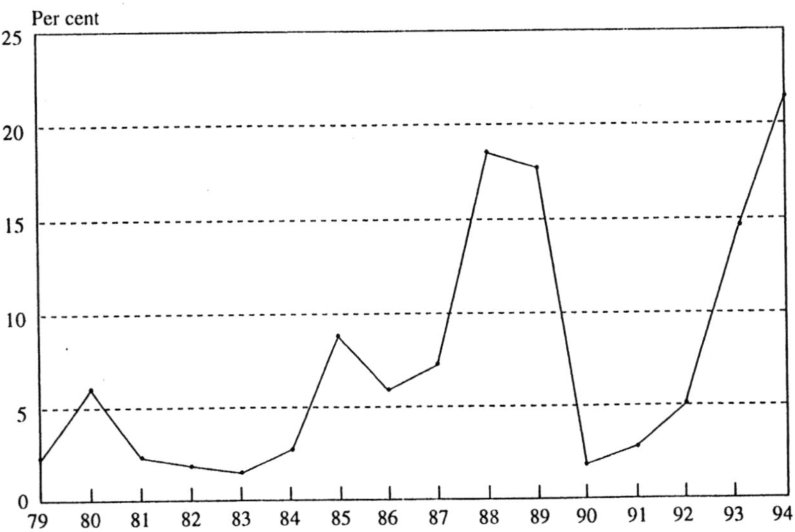

FIGURE 4.2 *Growth rates of retail price index, 1979–94.*

standard macroeconomic theory that excessive investment is only an indicator of economic overheating, rather than the cause. The question is, what caused the high growth of investment? As long as the central bank can effectively control money supply, where does excessive investment come from?

(2) Excessive consumption. Many argue that enterprises, particularly SOEs, tend to pay their workers excessive wages and in-kind benefits. According to this view, these enterprises find various ways to circumvent the government-imposed restrictions on wage and bonus growth; in many of the overheated periods, the growth rate of wages and bonuses exceeded the growth rate of productivity, thereby creating inflation. This argument, similar to that of excessive investment, points to a result of the economic overheating, rather than the cause. Again, if the money supply is strictly controlled, there would not be an excessive wage growth.

(3) Bad decision making of the central bank in terms of money supply. This 'bad decision' argument gets closer to the answer, but does the central bank understand that? If the central bank knows that a high growth of money supply causes high inflation, why has it repeatedly failed to control the money supply at the desired level?

My hypothesis is that the central bank's money supply decision is largely endogenous given the current institutional set-up in China. Although a modest monetary growth may be optimal to the central bank in an ideal environment, it is not so when it faces various institutional constraints inherent in the system. In other words, the central bank itself should not be blamed for the excessive monetary growth we observed over the past years; rather, the suboptimal result is an equilibrium outcome of the Chinese monetary institutions. Without changing certain institutional arrangements, money supply-generated economic instability will be likely to continue. The next section presents a theory that explains how excessive money growth is generated in a central–local monetary game.

WHY CANNOT THE CENTRAL BANK CONTROL MONEY SUPPLY AT THE DESIRED LEVEL?

As mentioned above, the central bank may well understand that excessive money growth leads to high inflation, so it does plan to control the total money supply at a desirable level. This intention can be seen from the comparison of the planned total credit target and the actual credit supply, as well as the comparison of planned cash issuance and the actual cash issuance. In many years, except those of recession, the planned targets were well below the actual figures (See Tables 4.1 and 4.2). In other words, the pre-announced credit targets and cash issuance targets were repeatedly revised upwards during the course of implementation. This indicates that although the central

TABLE 4.1 *Comparison of actual credit supply and planned credit supply (in billions of yuan)*

Year	*Planned target (a)*	*Actual volume (b)*	*(a)/(b)*
1983	35.4	37.8	1.07
1984	42.3	98.8	2.34
1985	71.5	148.6	2.09
1986	95.0	168.5	1.77
1987	122.5	144.2	1.17
1988	–	151.8	–
1989	–	185.1	–
1990	170.0	275.7	1.60
1991	210.0	287.8	1.37
1992	280.0	386.4	1.38

SOURCE Fan *et al.* (1993).

TABLE 4.2 *Comparison of actual currency issued and planned target (in billions of yuan)*

Year	*Planned target (a)*	*Actual volume (b)*	*(a)/(b)*
1983	6.0	9.1	1.51
1984	8.0	26.2	3.28
1985	15.0	19.6	1.30
1986	20.0	23.1	1.15
1987	23.0	23.6	1.03
1988	20.0	68.0	3.40
1989	40.0	21.0	0.53
1990	40.0	30.0	0.75
1991	50.0	53.5	1.07
1992	60.0	115.8	1.93

SOURCE Fan *et al.* (1993).

bank knows that a modest growth rate of credit and cash supply is desirable when it makes the annual plan, there exists some mechanism that induces (or forces) the central bank to deviate from its pre-announced money supply decision.

We will argue that the credit ceilings are not respected because of a 'monetary game' being played between the local banks (local branches of the specialized banks) that represent the interests of the localities and the PBC (see Chapter 5 for a formal treatment of this game). The structure of the central–local monetary game is as follows. There are two game players: a central bank and a representative local bank. The central bank's objective includes two elements: controlling inflation and increasing total output. The local bank, which is controlled by a local government, maximizes the market value of local output and ignores the impact of its credit allocation on the price level. Under a decentralized banking system, the control variable of the central bank is the credit ceiling, which defines the total amount of credit that the local bank can extend during a year. Given this credit ceiling, the local bank decides how to allocate the credit ceiling between two types of projects: the high yield projects and the essential projects. Examples of high yield projects include those in the highly profitable manufacturing sectors and examples of essential projects include those in the agriculture, infrastructure, and energy sectors that are less than profitable. In China, lendings to these two types of projects are often referred to as commercial lending and policy lending, respectively.

Credit allocation between the two types of projects is the main instrument that the local bank uses in playing games with the central bank. The local bank chooses investment projects based on market values (that is, it favors high yield projects rather than essential projects). Unlike the local bank, the central bank values different projects using shadow prices, so it tends to favor the essential projects more than the local bank does. Knowing that the central bank values certain essential projects and has the flexibility to revise the credit ceiling *ex post,* the local bank purposely under-invests in the essential projects and creates pressures on the central bank to grant additional credit allowance. After observing that the local bank has under-financed the essential projects, the central bank has to provide additional credit to the essential project, that is, to revise the credit ceiling upwards.

It is rigorously shown in Chapter 5 that if the center can commit to the pre-announced credit ceiling, the local government will restrain itself from distorting the investment structure. The reasoning is that doing so – distorting the investment structure – would not induce additional credit from the central bank. Therefore, under a system of the center's commitment, the total credit supply and inflation will be lower compared to the case without the center's commitment. The interesting question is, knowing that a commitment to the pre-announced credit plan can improve the outcome, why does the center not commit? This is because after the local bank has allocated the initial credit ceiling in the distorted way, the central bank always finds that it is in its own interest to grant additional credit to correct the distorted investment structure. Without some externally imposed restriction that prohibits the central bank from revising credit ceilings, the only equilibrium of the game is 'credit ceiling revision, high credit supply, and high inflation'.

There are many anecdotal stories that reflect the spirit of this game. The famous story of 'white notes' is an example. By the end of every year, many provinces reported to the central bank that they had used up all the credit allowance and had no money to purchase grain. They had to issue 'white notes' – IOUs – to the farmers in exchange for grain. This practice caused strong resentment from the farmers. To redeem the IOUs, the center was forced to offer additional credit to these regions. By the end of 1992, there were several accounts of regional unrest among farmers in a number of provinces, partly due to local governments' 'white notes' issuance. This worried the central government a great deal and the State Council vowed to never allow the local governments to issue white notes again. Unfortunately, recently there were new accounts reporting that such a practice recurred in 1993's grain procurement.[6]

In 1992 and 1993, in addition to the 'monetary game' described above, a new problem emerged and further contributed to the excessive monetary growth. Many specialized banks found ways of funneling funds out of the state banking system and, thus, avoided the ceilings altogether. In 1992, the growth rate of credit channeled through the state banking system was 20 per cent, while M2 (including cash, demand deposits, and time deposits) growth reached 31 per cent mainly because the growth of lending by non-bank financial institutions (NBFIs – including trust and investment companies, urban credit cooperatives, rural credit cooperatives, finance companies, and so on) went unchecked. It seems that while reporting that the specialized banks were lending to the interbank market, they in fact channeled substantial funds to enterprises through NBFIs. This problem was a result of the central bank's failure to enforce its regulation that NBFIs are not allowed to deal with savings and loans.

REFORM CYCLES AND MONETARY CYCLES

Over the past decade, along with the monetary cycles, China also experienced a number of reform cycles. A reform cycle can be roughly described as a process in which decentralization is followed by recentralization, and recentralization followed by another round of decentralization. Interestingly, the reform cycles have in general coincided with the monetary cycles: a monetary expansion was usually accompanied by a wave of decentralization, while a period of monetary contraction was also a period of recentralization. Unlike many China observers who believe that the political struggle between reformists and conservatives determines China's reform cycles, I propose here an economic explanation of why reform cycles occur and why they coincide with the monetary cycles.

Under stable macroeconomic conditions (low inflation), the central government's main concern is how to increase the economic growth rate. Deng Xiaoping's call for rapid development in 1992 was a recent example. The intention for rapid development is quickly translated into decentralization, which increases the localities' and enterprises' autonomy as well as the incentive to expand local output. The decentralization measures include rendering more fiscal power to the localities, giving more authority to localities in project approval, credit and material allocation, and letting firms retain high shares of profits, reducing restrictions on firms' pricing and wage setting, giving more firms importing and exporting rights, and so on. On the one hand, with relatively better information at the local level, the localities can select and approve projects more efficiently than the center. Probably the management and operation of local projects are also more efficiently conducted by local officials that by centrally designated ones. On the other hand, however, with the existence of the structural distortions, these decentralization measures allow the localities to play the game with the center and, consequently, force the increases in credit and cash supply as well as government expenditure from the budget system (the same logic as credit plan revision). The increase in money supply inevitably leads to high inflation. Once faced with high inflation, the central government's first priority shifts from economic growth to maintaining economic and political stability. Because of the slow functioning or simply the non-existence of indirect macrocontrol measures (such as the discount rate, open market operations, and reserve ratios), the most convenient and efficient option for stopping inflation is to recentralize powers that had been delegated to the localities and the firms, including the power to allocate credit among projects.

This 'monetary cycle led reform cycle' theory is confirmed by many accounts during the last few years. The 1987 expansion was accompanied by significant decentralization in most of the areas mentioned above. When inflation surged to 18.7 per cent in 1988, the central government decided to reimpose controls over many freed commodity prices, to recentralize credit allocation rights, to force the local governments to cut fixed assets investment by 20 per cent uniformly, and to freeze wages in the state- and collectively owned enterprises. As a result, inflation was brought down to 2 and 3 per cent in 1990 and 1991, respectively, but GNP growth also dropped to 4 per cent in 1990 and picked up only moderately to 7 per cent in 1991. As inflation seemed not to be a pressing concern, 1992 witnessed another round of decentralization, during which the central government allowed most provinces to render the provincial level project approval rights to the county level governments. A large number of commodity prices were also freed from government control and firms were given more leeway in determining wages and bonuses. Supervision over banks' operations was also loosened and, consequently, local banks poured a large amount of money into the estate and stock markets. When the retail price index again rose to nearly 20 per cent in 35 major cities during the first 6 months of 1993, the central government decided to launch an austerity program. While some of the policies were indirect measures,[7] many of them were still of administrative recentralization. These administrative measures included the following: requiring the local governments to sell their allotted state treasury bonds by 15 July, ordering banks to recall loans and ensuring they did not surpass credit ceilings limits set by the central bank, freezing most automobile imports for the next half year, dispatching work parties from Beijing to ensure the measures were being carried out, and ordering banks to retain sufficient funds to pay for the year's agricultural harvest. By the end of 1993, the government reportedly announced measures to recontrol wage payments and the prices of 17 major commodities.[8]

The implication of the above analysis is that without changing the institutional features of the current monetary system, the monetary cycles will be an inevitable phenomenon and, consequently, reform cycles will continue to occur. Reform of the current central–local monetary relations is urgently needed if the monetary system is to play a stabilizing role.

MONETARY REFORM IN 1994

In late 1993, in an attempt to establish a well functioning monetary control system, the central government announced a reform plan to begin in 1994. The main points of this reform plan were as follows.

(1) The central bank and its functions. The central bank's main function would be to formulate and implement monetary policy. 'The primary objective of the monetary policy is to maintain the stability of the value of the currency, and on that basis, to support the growth of the economy'

(FBIS, 1993). Other functions of the central bank would include enforcing strict supervision over financial institutions, conducting payment clearing and settlement, and issuing banknotes.

(2) The central bank's relations with specialized banks and other commercial banks. According to the plan, the authority to formulate monetary policies must be concentrated in the hands of the PBC head office. To strengthen the central bank's hand in this task, powers previously given to PBC branches to control funds would be taken back to the bank's headquarters 'to lessen the influence and interference of local governments in the central bank's macro-financial control' (FBIS, 1993). In particular, a system consisting of only six regional PBC branches would be established to replace the old system in which each PBC branch is located in a province. The heads of the regional PBC branches would be directly appointed by the PBC headquarters without consulting the provincial governments.

(3) Instruments to conduct monetary policies. The PBC would gradually use more indirect means to control the money supply, such as open market operations, discount rate adjustments, reserve ratios, and operations in the foreign exchange market. The application of mandatory credit plan would be gradually reduced.

(4) Deficit financing. In the new system, deficit financing would mainly rely on selling bonds, while central bank financing through printing money would be prohibited. This reform was referred to as 'increasing the central bank's independence' (FBIS, 1993).

(5) Policy lending and commercial lending. Commercial lending and policy lending would be separated. The four specialized banks would be transformed into commercial banks that aim at profit making while taking the risks themselves. Competition among the commercial banks would be encouraged by allowing the specialized banks to conduct businesses formerly outside their designated scope. Investment, securities, and saving deposits institutions would be separated from the commercial banks.

The Agricultural Bank, the Bank of China, the People's Construction Bank, and the Industrial and Commercial Bank would become independent commercial banks, the Communications Bank of China and some other banks would be turned into joint stock banks, and localities would be encouraged to set up their own banks. The existing urban and rural credit cooperatives would be transformed into cooperative banks. Three policy lending banks – the State Development Bank, the Import–Export Bank, and the Agricultural Development

Bank – would be set up. These policy banks would grant policy loans to selected projects in accordance with state industrial policy. The capital source of these banks would include budgetary transfer, social insurance, postal deposits, and investment funds.

(6) Interbank market lending. The maximum length of maturity allowed for interbank loans would be shortened from 3 months to 3 days.

ASSESSMENT OF THE 1994 MONETARY REFORM

The 1994 reform plan has addressed a number of important drawbacks in the old system that led to the central bank's weak control over money supply. In particular, the proposed establishment of the new PBC branch system, the centralization of personnel management of the PBC branches and specialized banks, and the separation of policy lending from commercial lending are all in the right direction and are very much in line with our previous analyses. These measures will, to some extent, reduce the local governments' ability to bargain with the central bank for excessive credit supply. The central bank's independence from the Ministry of Finance (MOF) will prevent easy monetarization of fiscal deficits, thus contributing to a stable money supply.

However, several problems that were responsible for the excessive monetary growth in the past are still left unaddressed. First, the central bank is still under the strict control of the State Council. It is often the case that for various political reasons, the State Council is more 'radical' (in the sense that it favors rapid economic growth rather than fighting inflation) than the central bank, if the central bank is given the authority to express its own view. This distinction is confirmed from my interviews with several PBC officials. Under the State Council, the central bank is not likely to be able to commit to a prudent money supply (in the form of credit ceiling or reserve ratio) since the State Council is often unable to resist demands such as rescuing SOEs, maintaining 'political stability', or solving bottle-neck problems. Without such a commitment, the fundamental source of the monetary game still exists and excessive money growth may recur in other forms under the new system.

Under pressures from local governments and/or enterprises, however, a central bank that is more conservative than the central government is less likely to be tempted to grant additional loans or increase cash issuance targets. This suggests a central bank that is independent from the central government. A number of cross-country studies have presented empirical evidence suggesting that a more independent central bank is usually associated with a lower inflation rate (Alesina and Summers, 1993). We recommend that China places the central bank under the Congress, and grants the central bank independence in the following sense: its decisions are not reviewed by other bodies except the Congress or its standing committee, its governor is appointed for extended periods and cannot be removed without cause, it does not have to seek approval for its budget, and it is not subject to an external auditor. Unlike the government proposed 'central bank independence' – independence from other ministries – which is only an administrative order and can be easily revoked by the central government itself, the independence that we propose must be written into the constitution and be protected by the legal system.

Second, the fundamental source of the problem of excessive money growth – structural distortions – is not given enough attention. With the existing price distortions and the administrative and financial entry barriers to certain key sectors, the problem of investment distortion will continue to create pressures on the central bank for money supply. Although in the new system the regional PBC branches will no longer represent the interests of the localities, local governments may find alternative ways to persuade the center for allowing a high growth of money supply. Therefore, continuing price reform, phasing out state monopoly in sectors such as energy, electricity, telecommunications, and transportation, and fostering a more efficient financial market that can facilitate large-scale investment should go hand in hand with the monetary reform.

Finally, the reform plan does not mention eliminating interest rate control. In the presence of interest rate control, there always exist incentives for the commercial banks to divert funds from the banking system to black market lending. Problems similar to the illegal lending through the interbank market in 1992–3 will be difficult to eradicate without interest rate decontrol. Other benefits of liberalizing interest rates include an increased investment rate and more competition among banks that can help improve efficiency. Nevertheless, strengthened central bank supervision is a prerequisite for interest rate decontrol.

CONCLUSIONS

This chapter argues that the central bank *per se* is not to be blamed for the excessive monetary growth that created repeated high inflation; rather, the institutional features of the current central–local monetary relations generate this result endogenously. A central–local monetary game is proposed as a theory to explain how local governments take advantage of the central bank's lack of commitment to a pre-announced credit policy by forcing the central bank to revise credit ceilings upwards, thus creating inflation. This game also explains how reform cycles – the alternating of decentralization and recentralization – over the past decade could be a consequence of the central–local monetary interaction and why reform cycles have coincided with the monetary cycles.

In the light of central–local monetary relations, this chapter examines the implications of the 1994 monetary reform. This reform plan has addressed a number of important drawbacks of the old system that led to the central bank's weak control over money supply. In particular, the proposed establishment of the new central bank branch system, the centralization of personnel management of the central bank branches and specialized banks, and the separation of policy lending from commercial lending are all in the right direction. Nevertheless, further reforms towards a more independent central bank, the application of indirect measures to exercise monetary control, and the elimination of various structural distortions that led to the conflict of interest between the center and the localities will be necessary for an effective monetary management system in the 1990s.

5 Central Government Credibility and Monetary Expansion: a Game Theoretic Model

INTRODUCTION

This chapter formalizes the central–local monetary game, discussed in the previous chapter, in a formal model. This model shows how China's inflation over the past decade may be a consequence of the decentralized financial system in which the central bank determines the credit ceilings and the local banks allocate credits between different projects. The institutional factors that contribute to inflation will be identified and particular attention will be given to the role played by the central bank's lack of commitment to its credit policy.

The first section reviews the literature on China's inflation and the relationship between my model and those in the literature. The second section describes the structure of the credit allocation system used in the model. In the third and fourth sections, the credit game between the central bank (center) and a local bank (a representative region) under a decentralized credit allocation system is discussed in detail. The third section considers the case where the center commits not to revise the credit ceiling and the fourth section considers the case where such a commitment cannot be made. The equilibrium outcomes of these two cases are compared in the fifth section. It is shown that the no commitment regime yields a higher credit supply, a higher inflation rate, and a lower welfare level, compared to the commitment regime. Conclusions are offered in the last section.

LITERATURE ON CHINA'S INFLATION

A number of recent articles have discussed China's money supply and inflation. Chow (1987) applied the quantity theory of money to China and found that the ratio of money supply to real output is an important variable in explaining the price level. Feltenstein and Ziba (1987) estimated a money supply function and argued that the change in money supply (broad money) can be explained by changes in wage payments, agricultural procurement payments, and government deficits. Feltenstein and Ha (1989) estimated the repressed inflation and liquidity overhang. Woo *et al.* (1993) argued that increasing government subsidies to loss-making, state-owned enterprises is responsible for increasing deficit and macroeconomic instability. Garnaut and Ma (1993) and Cardoso and Yusuf (1994) provided some descriptive analyses based on the recent experience of China's inflation. While these studies confirm that an excessive money supply would create inflation and that the source of money supply could be decomposed into a number of factors, they do not answer the question of why the central government could not or was not willing to control the money supply at a level that yielded a stable price level.

Some Chinese scholars propose that inflation arises during reform for 'frictional' reasons (see, for example, Fan, 1989). During the decades under planning, the structure of production was mainly determined by the planners' demand for producer and investment goods. Reform in the 1980s allowed a rapid increase in consumer demand and a greater proportion of national income has been put in the hands of consumers. Demand for consumer goods rises faster than the supply, as the adjustment of the production structure takes time. The lag between the change in demand and that in supply results in 'frictional inflation'. Again, this theory does not tell us why the central bank simply accommodates the high demand for money. In addition, the literature in Chinese largely remains at the descriptive and institutional stage, mainly due to the lack of effective modeling tools.

Among the published literature, Sicular (1990) and Wang (1991) provided two formal theoretical models that describe the process of inflation creation in China. Sicular (1990) proposed that inflation can arise due to inherent contradictions between plan and market. Her model uses China's grain market as an example of the two-track system, under which the government requires farmers to sell certain quantities of their grain output to the state at a below-market planned price and then the government sells these products at a low, ration price to urban consumers. The model shows that market opportunities (that is, the difference between market price and planned price) encourage plan evasion and cause a shortfall in planned deliveries. This forces the government to either raise the planned procurement price or purchase more from the free market. Both measures imply increased government subsidies, which, under the assumption that subsidies are covered by money printing, generates inflation.

While increased government subsidies in the process of grain procurement have surely contributed to an increasing government deficit and the resulting inflationary pressure, for several reasons it is not clear how significant this mechanism is to the overall price change. First, an increase in price subsidy may or may not increase government deficits proportionally, depending on the overall budgetary arrangement. Second, government deficits in China are not totally financed by borrowing from the central bank. In recent years, more than 70 per cent of the deficits have been financed by government debts. A simple ordinary least square (OLS) regression of government borrowing from the central bank on price subsidies yields an R^2 of only 0.28, indicating little explanatory power of the independent variable.[1] Third, one cannot assume that 1 yuan of government borrowing from the central bank means that the central bank creates an additional yuan of base money. To what extent government borrowing from the central bank is financed by money creation depends on the changes of all other items in the central bank's balance sheet. In addition, the accommodative policies of the government (for example, raising procurement price or increasing subsidies to the urban consumers) are assumed in the model; it does not explain why the government chooses to do so.

Wang (1991) based his theory on decentralized resource control at the firm's level. His model assumes that each firm maximizes its output and the government maximizes the total output of all firms. The firms allocate their own resources between fixed capital investment and circulating funds, while the government only provides the circulating funds to the firms. To compete for resources from the government, the firms allocate less than the optimal amount of resources to circulating funds. The prisoner's dilemma game among firms leads to a lower level of total output than the optimum. To achieve a high level of output, the government decides to print money to provide more circulating funds to the firms, thus creating inflation.

My model presented in this chapter is similar to Wang (1991) in the sense that both attempt to link resource misallocation at the firm's level (Wang's model) or at the local level (my model) to the government's decision to create money (Wang's model) or to the central bank's decision to allow a higher credit supply (my model). However, there are a number of important differences between the two models. First, Wang (1991) stressed the role of resource misallocation between circulating funds and fixed investment within firms in the process of inflation. My model emphasizes the structural distortion between different types of investment projects (for example, agricultural versus manufacturing) due to local government influence. Second, Wang (1991) assumed that the government's objective is to maximize output only, with no concern about inflation. This objective, which is inconsistent with the fact that the Chinese leaders have regarded inflation as a life-threatening destabilizing factor, naturally generates the bias towards high inflation. My model, following Barro and Gordon (1983), assumes that the central bank cares about both output and inflation control. Third, in

Wang's (1991) model, resource misallocation is generated in a game of prisoner's dilemma among firms. Each firm misallocates its own resources at the expense of other firms, and the source of the inflation is interfirm externality. My model is a game played between the central bank and a representative local bank, where the central bank's lack of commitment to a fixed credit plan induces investment distortion, which in turn creates pressure on the central bank for a higher level of credit. In other words, the central bank's time-inconsistent behavior with regard to credit supply is the source of inflation. Nevertheless, my model does not 'disprove' Wang's (1991) theory; rather, the two models are complementary in the sense that each stresses one aspect of the causes of inflation in China.

Over the recent high-growth–high-inflation period, evidence from China tends to substantiate my story based on investment distortion between different types of projects. The famous story of 'white notes' is the best example of how local governments distort the investment structure in order to force the central bank to increase the credit ceiling. By the end of every year over the past few years, many provinces reported to the central bank that they had used up all the credit allowance and had no money to purchase grain. They had to issue 'white notes' – IOUs – to the farmers and caused strong resentment from them. To redeem the IOUs, the center was forced to offer additional credit to these regions (Li and Jiang, 1993). In the first quarter of 1993 alone, the local branches of the specialized banks lent a net of 100 billion yuan to non-bank financial institutions for operations including acquisition of real estate assets and purchase of securities (World Bank, 1994a), while leaving many essential projects in the central government's priority list under-financed. This became one of the primary reasons for the dismissal of the central bank governor Li Guixian in June 1993.

In terms of policy implications, Wang (1991) had no discussion on the normative elements of his model. In contrast, my model shows the theoretical basis for a number of institutional changes proposed in China's 1994 reform package, including the separation of policy lending from commercial lending and the reorganization of central bank branches from a province-based to a region-based system. In addition, I offer a case for central bank independence based on the finding of the model, which has not been formally discussed in China.

DESCRIPTION OF THE CREDIT ALLOCATION SYSTEM

This section presents the model's basic assumptions about the environment of the economy. The economy consists of a central bank and a representative local bank. We start by highlighting a number of important institutional assumptions which will have significant implications in policy discussions. All of these assumptions are based on facts of the current Chinese system.

(1) The local bank represents the local government (the region). Under the current decentralized Chinese system, the local banks are in effect controlled by the local governments.
(2) Under a decentralized banking system, the local bank has the authority to allocate credit across projects within its jurisdiction.
(3) As the basic means of controlling inflation, the center determines a credit ceiling for each region. The region is not allowed to make loans exceeding the ceiling, but the center can revise the credit ceiling.

There are two types of project in the economy. The first type is called the high yield project (project 1). In China, the processing and service industries are an example of this type of project. The second type is called the essential project (project 2). Examples of the second type of project include those in the agricultural, infrastructure, and energy sectors.

(4) Project 1 has a higher monetary yield than project 2. This reflects the fact that the projects in most processing and service sectors are usually more profitable than those in the agricultural, infrastructure, and energy sectors in China.

The most important reason for the profitability differential between high yield projects and essential projects is the government's asymmetric price control. While the government has eliminated most of the price controls in the light industry and service sector, it still controls a large part of the output prices in the agricultural, infrastructure, and energy sectors. For example, the profit rate of the entire coal mining and production sector has been negative for nearly a decade due to price control.[2] For the same reason, approximately one-half of the loss-making enterprises are located in the energy sector. In the agricultural sector, the government still maintains the planned procurement prices that are lower than market prices. Prices for major transportation and telecommunications services are still directly set by the central ministries. These remaining price controls are difficult to liberalize in the short run mainly due to political constraints.[3] In addition to price control, another important reason for the essential projects' inability to attract enough financial resources is that they tend to generate positive interregional externalities.

Given the above assumptions, we can construct models for the center and the region's behaviors. The center values output but dislikes inflation. This assumption is used widely in the monetary policy literature.[4] We also assume that the center values the two projects equally in its economic planning.[5] The center's objective is therefore to maximize the following welfare function:

$$U = f(L_1) + f(L_2) - \Pi(L_1, L_2) \tag{5.1}$$

where L_1 is the credit extended to project 1 and L_2 is the credit extended to project 2. We assume that $f(L_1)$ and $f(L_2)$ represent the real outputs of the two projects. For simplicity, the two projects have the same physical production function, $f(.)$, with the properties of $f'(.) > 0$, $f''(.) < 0$, and $f'(L) \to \infty$ as $L \to 0$.[6] $\Pi(L_1, L_2)$, the inflation rate, is a function of outstanding loans. The center's choice variable varies in different cases. Under a centralized system, the center directly chooses L_1 and L_2, that is it allocates credit across projects. Under a decentralized system, as is the case in post-reform China, the center chooses a total credit ceiling $L(= L_1 + L_2)$ and lets the region allocate L between projects.

To keep the analysis tractable, we suppose that investment and inflation take place in two distinct periods: the base period and the current period. Investment is made in the base period when the aggregate price level is fixed at 1. The resulting inflation only occurs in the current period. This assumption allows us to use the nominal inputs L_1 and L_2 in the production functions.

The inflation rate is simply derived from the following equation:

$$PY = M \tag{5.2}$$

where P is the aggregate price level and M is the total money supply. $Y = f(L_1) + f(L_2)$ is the total real output. The velocity is assumed to equal unity. Since the base period's aggregate price level equals 1, the inflation rate is

$$\Pi = P - 1 = M/Y - 1. \tag{5.3}$$

To simplify the analysis, we also assume that the money supply consists only of credit, so $M = L$. This is a reasonable assumption when total currency is small relative to outstanding loans in the economy.[7] Equation (5.3) becomes

$$\Pi = (L_1 + L_2)/[f(L_1) + f(L_2)] - 1. \tag{5.4}$$

The region, unlike the center, is concerned with the market value of local output rather than with national welfare. It responds to the actual prices which undervalue project 2 (from the center's point of view). In addition, since the total number of regions is large, the representative region does not take into account the effect of its choice variable on the national inflation rate.[8] The region determines the allocation of L_1 and L_2 to maximize its objective:[9]

$$V = f(L_1) + \delta f(L_2) \tag{5.5}$$

subject to the constraint

$$L = L_1 + L_2 \tag{5.6}$$

where L is the credit ceiling given by the center. The coefficient δ $(0 < \delta < 1)$ in front of project 2's production function captures the region's undervaluation of project 2.[10] In this model, we assume that the region's undervaluation of project 2 is a result of government price control over project 2.

We assume that the banking system is a decentralized one, in which the center only determines a total credit ceiling and the region allocates credit between projects with the given credit ceiling. This set-up mimics the actual system adopted by China since 1983. With the authority to allocate credit within a certain ceiling, the region has the instrument to play a game with the center. There are two possible regimes, under which the game will yield very different results. In the first, the center is able to commit not to revise the credit ceiling upward after the initial ceiling is announced. The game involves two stages: in the first stage the center chooses L and in the second stage the region chooses L_1 and L_2. However, the center's commitment is not credible since the center can achieve a higher utility level by reoptimizing *ex post*. In the second regime, the center does not commit to any pre-announced credit ceiling, that is it is free to provide additional credit to project 2 after the region allocates the credit. Knowing that the center's pre-announced credit ceiling will be revised, the region will react by choosing a credit allocation that is different from that under the commitment regime. The second regime is inspired by the fact of repeated credit ceiling revisions in the current Chinese credit allocation system.

We will show that under the no commitment regime, the equilibrium yields a higher level of total credit, a higher inflation rate, and a lower welfare level, than those under the commitment regime. This suboptimal equilibrium, we believe, well describes the functioning of the current Chinese decentralized banking system. This comparison will have important policy implications for future reforms in the Chinese banking system.

The next section discusses the equilibrium under the commitment regime and the fourth section discusses the equilibrium under the no commitment regime. A comparison of the major results under the two regimes is made in the fifth section.

THE COMMITMENT REGIME

Under the commitment regime where no revision of the credit ceiling is involved, the game has two stages. In the first stage, the center determines the optimal credit ceiling, given the region's reaction function to the credit ceiling. In the second stage, given the credit ceiling, the region chooses how to allocate the credit between two projects. This is a Stackelberg game in which the center acts as a leader. The Stackelberg equilibrium is defined by two conditions: the center's credit ceiling is optimal, given the reaction function of the region; and the region's credit allocation is optimal, given the center's credit ceiling. We assume that the center knows the region's utility function. The region also has perfect information about the center's objective and is informed about the center's initial credit ceiling when making its decision.

The problem can be solved backward. First, the region optimizes with a given credit ceiling. That is, it solves for the optimal credit allocation as a function of the credit ceiling. Then, the center solves for the optimal credit ceiling, taking into account the region's reaction function.

The region maximizes its objective

$$V = f(\beta L) + \delta f((1 - \beta)L) \tag{5.7}$$

by choosing the share of credit to project 1, β. δ is the region's discount factor on project 2. We restrict $\beta \leq 1$, that is, the region is not allowed to allocate more than 100 per cent of the total credit to a project. Given the assumption that $f'(L) \to \infty$ as $L \to 0$, it is also easy to see that the optimal β is greater than 0. The first-order condition equates the values of marginal product of project 1 and project 2:

$$f'(\beta L) = \delta f'((1 - \beta)L). \tag{5.8}$$

It is easy to see that, since $f''(.) < 0$, the second-order condition holds globally. Using equation (5.8) and the production function $f(L) = L^{\alpha}$, we obtain the optimal β:

$$\beta = 1/(\delta^{1/(1-\alpha)} + 1). \tag{5.9}$$

From the assumption $0 < \delta < 1$, we have $\beta > 0.5$.

Remark 1: under the decentralized system with the center's commitment, project 1 receives a higher share of credit ($\beta > 0.5$) than project 2 ($1-\beta < 0.5$). From the center's point of view, project 1 is over-financed and project 2 is under-financed.[11] The more the region undervalues project 2 (the lower δ is), the smaller the share of credit $(1 - \beta)$ allocated to project 2 is.

The region's reaction function, equation (5.9), is taken into account by the center. Using the production function $f(L) = L^\alpha$ and equation (5.4), the center's objective function can be rewritten as

$$U = (\beta L)^\alpha + ((1-\beta)L)^\alpha - L^{1-\alpha}/[\beta^\alpha + (1-\beta)^\alpha] + 1.$$

The first-order condition determining L is

$$\alpha[\beta^\alpha + (1-\beta)^\alpha]L^{\alpha-1} - (1-\alpha)L^{-\alpha}/[\beta^\alpha + (1-\beta)^\alpha] = 0,$$

which yields the optimal credit ceiling

$$L^c = \{[(1-\alpha)/\alpha]/[\beta^\alpha + (1-\beta)^\alpha]^2\}^{1/(2\alpha-1)}. \qquad (5.10)$$

We can verify that L^c is the global maximum given the assumption of $\alpha < 1/2$.[12] In the equilibrium, the credit given to project 1 is

$$L_1^c = \beta\{[(1-\alpha)/\alpha]/[\beta^\alpha + (1-\beta)^\alpha]^2\}^{1/(2\alpha-1)}$$

and the credit given to project 2 is

$$L_2^c = (1-\beta)\{[(1-\alpha)/\alpha]/[\beta^\alpha + (1-\beta)^\alpha]^2\}^{1/(2\alpha-1)}$$

where the superscript c on L, L_1 and L_2 stands for commitment. The total output, the inflation rate, and the center's utility level can be calculated accordingly.

THE NO COMMITMENT REGIME

Now consider the more realistic case where the center can increase the credit ceiling after the initial ceiling is announced, as manifested repeatedly by China's experience. The task of this section is to show that given the flexibility to increase the credit ceiling, the center has the incentive to do so. In other words, the center's commitment not to revise the credit ceiling is not credible. We will also solve for the equilibrium for comparison with the commitment case.

The game proceeds as follows. In the first stage, the center determines the initial credit ceiling, knowing that the region will react to its choice and that an *ex post* increase in the credit ceiling is possible. The second stage is a Cournot game between the center and the region. In this stage, the region chooses how to allocate the credit between two projects, taking the center's additional credit as given. In the same stage, the center determines whether and how much additional credit should be provided to project 2, taking the region's biased credit allocation as given. The Cournot structure assumes that two parties move simultaneously or react to each other's move continously. This assumption is employed partially for mathematical simplicity, but it is also justified by the following evidence. In 1988 and 1992, the 2 years when the economy was overheated, the center made at least *two* to *three* major revisions and numerous minor revisions on many regions' credit ceilings, as it observed that many 'key' projects were under-financed. At the same time, observing the center's revision of the credit ceiling, the region continued to misallocate credits in order to impose more pressures on the center. Therefore, it is not realistic to assume a leader–follower (Stackelberg) game structure (for example, in stage 2, the region allocates credits; in stage 3, the center decides whether and how much to provide additional credits) between the region's credit allocation and the center's decision on additional credit. The real credit revision is not a once-for-all action but a learning-type adjustment process which leads to a Cournot model. This information structure is analogous to two firms' quantity adjustment game in industrial organization literature (Fudenberg and Tirole, 1991, p.23).[13]

First, we define the center's problem in the Cournot game in stage 2. The center's problem is to determine the optimal additional credit, taking the initial credit ceiling, L, and the region's credit allocation, β, as given. From the previous section, we know that the region under-finances project 2 (and over-finances project 1) from the center's point of view, so the center's natural reaction should be to consider a possible increase for project 2. The easiest way to formulate the problem is to let the center choose a percentage increase, a, in the credit ceiling to maximize the objective

$$U = f(\beta' L) + f((1-\beta' + a)L) - \Pi(\beta' L, (1-\beta' + a)L) \qquad (5.11)$$

subject to

$$a \geq 0$$

where $\beta' L$ is the credit extended to project 1, aL is the additional credit to project 2, and $(1-\beta' + a)L$ is the total credit to project 2. Since in stage 2 L is already determined, it is taken as given. β' is the share of credit going to project 1. The prime on β is to distinguish it from that under the commitment regime. The solution to this problem is an optimal a as a function of L and β'.

Next, we define the region's problem in stage 2. This region determines the optimal credit allocation, β', taking into account the initial credit ceiling, L, and the center's possible additional credit, aL, to project 2. The region chooses β' to maximize

$$V = f(\beta' L) + \delta f((1 - \beta' + a)L). \tag{5.12}$$

The solution to this problem is an optimal β' as a function of L and a.[14]

Third, we define the center's problem in the first stage, when it acts as a Stackelberg leader. The center chooses an initial credit ceiling, L, to maximize its objective

$$U = f(\beta' L) + f((1 - \beta' + a)L) - \Pi(\beta' L, (1 - \beta' + a)L). \tag{5.13}$$

When solving this problem, the center takes into account its own *ex post* reaction derived from equation (5.11) and the region's reaction function derived from equation (5.12).

The equilibrium is a set of strategy $\{L, a, \beta'\}$ that satisfies the following conditions.

(1) The center's initial credit ceiling is optimal given the region's credit allocation function (a function of the credit ceiling) and the center's choice of additional credit as a function of the initial credit ceiling, that is

$$L = \text{argmax } U(\beta'(L)L, (1 - \beta'(L)L) + a(L))L)$$

where $\beta'(L)$ and $a(L)$ are solved from the next two conditions.

(2) The region's credit allocation is optimal given the initial credit ceiling and the center's choice of additional credit, that is

$$B' = \text{argmax } V(\beta' L, (1 - \beta' + a)L))$$
$$\text{for given } L \text{ and } a.$$

(3) The center's additional credit is optimal given the region's credit allocation function and the initial credit ceiling, that is

$$a = \text{argmax } U(\beta' L, (1 - \beta' + a)L)$$
$$\text{for given } L \text{ and } \beta'.$$

The detailed derivation of the equilibrium credit ceiling (after revision) under the no commitment regime is given in Appendix 5.1. By comparing the equilibrium credit ceiling under the no commitment regime, L^{nc}, with the equilibrium credit ceiling under the commitment regime, L^{c}, we can answer the question of whether the center's optimal credit ceiling in the commitment case is *ex post* optimal.

Remark 2: the optimal credit ceiling in the commitment case, L^c, is less than the *ex post* optimal credit ceiling, L^{nc}, if the center is allowed not to commit. Therefore, if the initial credit ceiling is L^c, the center will find it *ex post* optimal to allow additional credit extended to project 2. The center's optimal initial credit ceiling from the commitment equilibrium is thus not credible.

By comparing the optimal credit allocation β' by the region under the no commitment regime, with the optimal credit allocation β by the region under the commitment regime, we have the following remark (see Appendix 5.1).

Remark 3: starting from the equilibrium in the commitment case (that is, given L^c as the initial credit ceiling) and knowing that *ex post* the center will allow additional credit to project 2, the region further distorts the investment structure by allocating a higher share of credit from the initial credit ceiling, L, to project 1 (that is, $\beta' > \beta$).

The intuition for a higher level of credit supply under the no commitment regime than that under the commitment regime is as follows. Under the no commitment regime, the region knows that the center has the *ex post* flexibility to provide additional credit to project 2. It therefore purposely keeps the proportion of its investment in project 1 at a level higher than that under the commitment regime ($\beta' > \beta$). The center thus finds it optimal to allow additional credit to project 2 in order to reduce the proportion of investment to project 1 from β' to β. We have shown that this is true in the case where the initial credit ceiling is L^c. In other words, even if the center wishes to act as under the commitment regime (that is, to announce that L^c is the final credit ceiling), as long as there is no mechanism that can make the region believe that such an announcement is credible, the region's best choice is to further distort the investment structure which leads to a higher credit ceiling than L^c.

COMPARISON OF EQUILIBRIA UNDER THE COMMITMENT AND NO COMMITMENT REGIMES

The previous section shows that the total credit ceiling under the no commitment regime is higher than that under the commitment regime. This section compares other variables in the two equilibria under the commitment and the no commitment regimes. We will show that the commitment regime will yield the same credit allocation structure (*ex post*), a lower output level, a lower inflation rate, and a higher welfare level when compared to the no commitment regime. The immediate implication of this comparison is that a mechanism that can help the center achieve commitment will be welfare improving.

Credit allocation

Equation (5.9), $\beta = 1/(\delta^{1/(1-\alpha)} + 1)$, gives the optimal share of credit going to project 1 under the commitment case. Equation (A5.6) (see Appendix 5.1), $L_1^{nc} = \beta' L = \phi\beta$, gives the *ex post* share of credit going to project 1 under the no commitment case. Therefore, under the two different regimes the shares of total credit going to project 1 are the same.

Remark 4: in equilibrium, the commitment and no commitment systems yield the same biased credit allocation with a share of β going to project 1. The *ex post* credit increase in the no commitment system does not solve the problem of structural imbalance existing in the commitment case.

Output

Because $L^{nc} > L^{c}$, we have

$$f(\beta L^{nc}) + f((1-\beta)L^{nc} > f(\beta L^{c}) + f((1-\beta)L^{c}).$$

Remark 5: the no commitment regime yields a higher real output level than that in the commitment regime.

Inflation

Inflation under the commitment regime is

$$\Pi^{c} = (L^{c})^{1-\alpha}/[\beta^{\alpha} + (1-\beta)^{\alpha}] - 1. \quad (5.14)$$

Inflation under the no commitment regime is

$$\Pi^{nc} = (L^{nc})^{1-\alpha}/[\beta^{\alpha} + (1-\beta)^{\alpha}] - 1. \quad (5.15)$$

For $\alpha < 1/2$ we have $L^{nc} > L^{c}$, which leads to

$$\Pi^{nc} > \Pi^{c}.$$

Remark 6: the no commitment regime yields in equilibrium a higher inflation rate, compared to the commitment regime.

Welfare

Welfare (the center's utility) under the commitment regime is

$$U^{c} = (\beta L^{c})^{\alpha} + ((1-\beta)L^{c})^{\alpha} - (L^{c})^{1-\alpha}/[\beta^{\alpha} + (1-\beta)^{\alpha}] + 1 \quad (5.16)$$

TABLE 5.1 *Simulation results of two credit allocation regimes*

	Commitment	*No commitment*
Initial credit ceiling	1.908	4.000
Initial loan to project 1	1.203	3.274
Initial loan to project 2	0.705	0.726
Additional loan to project 2	0.000	1.191
Ex post loan to project 1	1.203	3.274
Ex post loan to project 2	0.705	1.917
Total loan	1.908	5.191
Total output value	1.953	2.727
Inflation rate	–0.023	0.903
Center's utility	1.977	1.824

NOTE: $\alpha = 1/3$ and $\delta = 0.7$.

where L^c is the single maximizer of U^c:

$$L^c = \{[(1-\alpha)/\alpha]/[\beta^\alpha + (1-\beta)^\alpha]^2\}^{1/(2\alpha-1)}.$$

Welfare under the no commitment regime is

$$U^{nc} = (\phi\beta^\alpha + (\phi(1-\beta))^\alpha - \phi/[(\phi\beta)^\alpha + (\phi(1-\beta))^\alpha] + 1 \qquad (5.17)$$

if the initial credit ceiling is less than or equal to ϕ (see Appendix 5.1 for the derivation of ϕ).

It is easy to see that U^{nc} has the same functional form as U^c, only with L^c being replaced by ϕ. Since $\phi > L^c$ and L^c is the only maximizer of U^c, it is obvious that $U^{nc} < U^c$.

Remark 7: the no commitment regime yields in equilibrium a lower welfare level, compared to that under the commitment regime.

Table 5.1 presents a numerical example of the two equilibria for a particular set of parameters. The parameters used in the simulation are $\alpha = 1/3$ and $\delta = 0.7$. All the features mentioned in the previous remarks are confirmed by the simulation result.

CONCLUSIONS

This chapter develops a game theoretic model which shows how China's economic overheating over the past decade may be a consequence of its decentralized financial system. Under the decentralized system a game is played between the central bank (center) and the local banks (regions), which results in an equilibrium with inflation. A number of stylized facts from the decentralized Chinese economy are generated from the game. These facts are structural imbalance, high output level, and high inflation.

By analyzing the structure of the game, the institutional factors that are responsible for these stylized facts can be identified. Simply speaking, the conflict of interest between the central and local governments is the fundamental source of these problems. This conflict of interest arises because the center values the investment projects using shadow prices while the regions value the projects using market prices. Under a decentralized system, this conflict causes a biased investment structure and, consequently, imposes pressures on the center for granting additional credit to the under-financed 'essential' projects. Compared with the case where the center can centralize credit allocation without incurring any monitoring cost, the equilibrium under the decentralized system is suboptimal.

The lack of credibility of the center is the key factor that allows the translation of the conflict of interest between the center and regions into inflationary pressure. Since the center is unable to commit to the pre-announced credit ceiling, the regions strategically take advantage of this lack of commitment by distorting the investment structure even further in order to obtain additional credit from the center. The upward revision of the credit ceiling yields an equilibrium with a higher inflation rate and a lower welfare level than in the case where the center can commit to a fixed credit ceiling. If the center can convince the regions that it is able to commit, the regions' incentive to strategically distort the investment structure will disappear. Thus, the interesting question becomes 'Why does the center not want to commit when it knows that commitment will lead to a lower inflation and a higher welfare level?' It is shown in the model that the center's optimal policy – commitment to the pre-announced credit policy -- is time inconsistent (that is, not credible). Without a mechanism that restricts the center's policy change, the center will always find it optimal to revise credit policy when the investment structure is distorted.

The positive analysis of this model naturally gives rise to some normative implications regarding institutional reform. Recall the institutional characteristics we outlined in the second section. By changing these conditions (that is, changing the rules of the game), the equilibrium outcome will be very different. Carefully examining the roles of the assumptions results in four major policy recommendations.

(1) Accepting the fact that regions value projects differently from the center, increasing the center's cost of revising credit plans will help reduce the

regions' incentive to distort the investment structure and, hence, bring down the inflationary pressure. This cost increase can be realized through legislation which complicates the administrative procedures required for a revision of the credit plan. One option for the institutional arrangement is to place the central bank under the National People's Congress, instead of the State Council, to avoid constant pressures from the central government for policy changes.[15]

(2) Continuing price liberalization is one of the fundamental solutions to the problem of structural imbalance.[16] This will resolve the conflict of interest between the center and the regions caused by price control. Therefore, when macroeconomic conditions permit, price reform should be one of the government's priority tasks. Nevertheless, reasons other than price control, such as cross-regional externalities, may still cause different objectives between the center and the regions in selecting projects.

(3) In the presence of the conflict of interest between the center and the regions, recentralizing policy lending (lending to 'essential' projects) from the localities may be necessary. By doing so, the center can avoid the distortion that would have arisen had the credit allocation been made by the regions, although more financial burdens will be borne by the center.[17]

(4) Separating local banks from local governments. Under the current system, most local bank branches are subject to controls from local authorities who value only local output value. Under these pressures, the local banks are representatives of the local governments to request as high a credit ceiling as possible. This is one of the necessary conditions of our credit game. To reduce the influence of provinces on local PBC branches, two options can be considered.[18] First, direct supervision of the central bank over its local branches should be strengthened. Personnel arrangements, provision of necessary working facilities, and staff benefits of the local branches should all be the responsibility of the central bank. Secondly, the current system in which each province has its central bank branch should be changed into one where each central bank branch covers a few provinces.[19]

APPENDIX 5.1: DERIVING THE EQUILIBRIUM CREDIT LEVEL UNDER THE NO COMMITMENT REGIME

To obtain the equilibrium solution, we first solve for a and β' as functions of L and then substitute them into equation (5.13) to solve for the optimal L. For the time being we solve equation (5.11) without considering the constraint of $a \geq 0$. We will check the condition later. Using the production function $f(L) = L^\alpha$, equation (5.11) becomes

$$U = (\beta' L)^\alpha + ((1-\beta'+a)L)^\alpha - (1+a)L/[(\beta' L)^\alpha + ((1-\beta'+a)L)^\alpha]$$
$$= L^\alpha[\beta'^\alpha + (1-\beta'+a)^\alpha] - (1+a)L^{1-\alpha}/[\beta'^\alpha + (1-\beta'+a)^\alpha].$$

The first-order condition is

$$dU/da = \alpha(1-\beta'+a)^{\alpha-1}L^\alpha - L^{1-\alpha}\{\beta'^\alpha + (1-\beta'+a)^\alpha - (1+a)\alpha(1-\beta'+a)^{\alpha-1}\}/[\beta'^\alpha + (1-\beta'+a)^\alpha]^2 = 0. \quad \text{(A5.1)}$$

Equation (A5.1) implicitly defines a as a function of β and L.

Next, we solve the region's problem. Using $f(L) = L^\alpha$ and taking first-order condition from equation (5.12) we obtain

$$\beta' = (1+a)\beta \quad \text{(A5.2)}$$

where $\beta = 1/(\delta^{1/(1-\alpha)} + 1)$.

Solving equations (A5.1) and (A5.2) together (see Appendix 5.2), we obtain the optimal a:

$$a = \phi/L - 1 \quad \text{if } \phi > L$$
$$\neq 0 \quad \text{if } \phi \leq L \quad \text{(A5.3)}$$

where

$$\phi = \{(1-\beta)^{1-\alpha}\{[\beta^\alpha + (1-\beta)^\alpha] - \alpha(1-\beta)^{\alpha-1}\}/\alpha[\beta^\alpha + (1-\beta)^\alpha]^2\}^{1/(2\alpha-1)}.$$

Consider the case that the parameters are such that $a > 0$ (that is, $\phi > L$). Substituting equation (A5.2) back into equation (A5.1) gives the region's share of credit going to project 1:

$$\beta' = \phi\beta/L. \quad \text{(A5.4)}$$

From equation (A5.4), we also have

$$1 + a - \beta' = \phi(1-\beta)/L. \quad \text{(A5.5)}$$

The credit allocated to project 1 is

$$L_1^{nc} = \beta' L = \phi\beta \quad \text{(A5.6)}$$

and the credit allocated to project 2, including the additional credit, is

$$L_2^{nc} = (1 + a - \beta')L = \phi(1-\beta). \quad \text{(A5.7)}$$

Therefore the total credit to the two projects is

$$L^{nc} = L_1^{nc} + L_2^{nc} = \phi. \tag{A5.8}$$

Now we are in the position to consider whether the center has the *ex post* incentive to provide additional credit to project 2 if the initial situation is the commitment equilibrium. We compare the total credit under the no commitment regime (equation (A5.8)),

$$L^{nc} = \{(1-\beta)^{1-\alpha}\{[\beta^\alpha + (1-\beta)^\alpha] - \alpha(1-\beta)^{\alpha-1}\}/\alpha[\beta^\alpha + (1-\beta)^\alpha]^2\}^{1/(2\alpha-1)},$$

with the total credit under the commitment regime (equation (5.10)),

$$L^c = \{[(1-\alpha)/\alpha]/[\beta^\alpha + (1-\beta)^\alpha]^2\}^{1/(2\alpha-1)}.$$

Note that $\alpha < 1/2$ is required for the second-order condition to be satisfied. Shuffling through mathematics (see Appendix 5.3), we obtain

$$L^{nc} > L^c.$$

As long as the initial credit ceiling does not exceed the no commitment equilibrium (*ex post* optimal total credit), that is $L < \phi$, the center will provide additional credit

$$aL = \phi - L > 0$$

to project 2.

Now we can answer the question of whether the center's optimal credit ceiling in the commitment case is *ex post* optimal.

Remark 2: the optimal credit ceiling in the commitment case, L^c, is less than the *ex post* optimal $L^{nc}(= \phi)$ if the center is allowed not to commit. Therefore, if the initial credit ceiling is L^c, the center will find it *ex post* optimal to allow additional credit extended to project 2. The center's optimal initial credit ceiling from the commitment equilibrium is thus not credible.

Given remark 2 (that is, $aL > 0$), equation (5.15) states the following.

Remark 3: starting from the equilibrium in the commitment case, and knowing that *ex post* the center will allow additional credit to project 2, the region further distorts the investment structure by allocating a higher share of credit from the initial credit ceiling, L, to project 1 (that is, $\beta' > \beta$).

To see the center's optimal choice of initial credit ceiling L, we should solve the center's problem in stage 1. Substituting equations (A5.6) and (A5.7) into equation (5.7) yields

$$U^{nc} = (\phi\beta)^\alpha + (\phi(1-\beta))^\alpha - \phi/[(\phi\beta)^\alpha + (\phi(1-\beta))^\alpha] + 1$$

for all $L \leq \phi$. It is a constant for given parameters. Therefore, the center is indifferent to any initial credit ceiling as long as $L \leq \phi$. This is because for any L ($\leq \phi$) determined in the first stage, the center will choose an aL in the second stage to ensure that the *ex post* credit ceiling is ϕ.

APPENDIX 5.2: SOLVING FOR OPTIMAL ADDITIONAL CREDIT (*A*) UNDER THE NO COMMITMENT REGIME

Substituting equation (A5.2) into equation (A5.1), we obtain

$$\alpha[(1+a)(1-\beta)]^{\alpha-1}L^{\alpha} = L^{1-\alpha}\{[\beta(1+a)]^{\alpha} + [(1-\beta)(1+a)]^{\alpha} - \alpha(1+a)[(1-\beta)(1+a)]^{\alpha-1}\}/\{(1-a)^{\alpha}[\beta^{\alpha} + (1-\beta)^{\alpha}]\}^{2}$$

or

$$[L(1+a)]^{2\alpha-1}\alpha(1-\beta)^{\alpha-1} = \frac{[\beta^{\alpha} + (1-\beta)^{\alpha}] - \alpha(1-\beta)^{\alpha-1}}{[\beta^{\alpha}+(1-\beta)^{\alpha}]^{2}}.$$

Rearranging terms gives

$$a = \frac{\{(1-\beta)^{1-\alpha}[\beta^{\alpha}+(1-\beta)^{\alpha}]-\alpha(1-\beta)^{\alpha-1}]/\alpha[\beta^{\alpha}+(1-\beta)^{\alpha}]^{2}\}^{1/(2\alpha-1)}}{L} - 1.$$

Letting

$$\phi = \left\{\frac{(1-\beta)^{1-\alpha}\{[\beta^{\alpha}+(1-\beta)^{\alpha}]-\alpha(1-\beta)^{\alpha-1}}{\alpha[\beta^{\alpha}+(1-\beta)^{\alpha}]^{2}}\right\}$$

and using the constraint in equation (5.11) we have

$$\begin{aligned} a &= \phi/L - 1 \quad \text{if } \phi > L \\ &= 0 \qquad\qquad \text{if } \phi \leq L. \end{aligned}$$

APPENDIX 5.3: PROOF OF $L^{nc} > L^{c}$

From $\beta > 1 - \beta$, we have

$$(1-\beta)^{1-\alpha} < \beta^{1-\alpha}.$$

That is,

$$(1-\beta)^{1-\alpha}\beta^{\alpha}+1-\beta-\alpha<1-\alpha$$

or

$$\frac{(1-\beta)^{1-\alpha}\{[\beta^{\alpha}+(1-\beta)^{\alpha}]-\alpha(1-\beta)^{\alpha-1}\}}{\alpha[\beta^{\alpha}+(1-\beta)\alpha]^{2}}<\frac{[(1-\alpha)/\alpha]}{[\beta^{\alpha}+(1-\beta)^{\alpha}]^{2}}.$$

Therefore,

$$L^{nc}>L^{c}.$$

for $\alpha<1/2$

6 Intergovernmental Relations and Market Development

INTRODUCTION

Over the past 15 years, China has made tremendous progress towards deregulating its old planning system. A major form of such deregulation has been the decentralization of authority to lower levels of government in the areas of enterprise ownership, taxation, government spending, credit allocation, investment project approval rights, price and wage control, foreign trade management, the formation of industrial policies, and so on. While this form of deregulation has provided the localities with enormous incentives and autonomy in adopting policies for regional development, it has had a serious side-effect as local governments have tended to abuse their administrative and regulatory powers to restrict competition in order to protect the local economies. These activities have impeded commodity flows and enterprise mergers and acquisition across regions and sectors, and exposed business transactions to the risks of *ad hoc* local government intervention.

Economic theories clearly make the case that the rules of the market critically affect the efficiency of resource allocation and the distribution of wealth. Without legal protection, the cost of market transactions will be very high. It is clear that as the Chinese central government withdraws from direct participation in many realms of the economy, it must engage itself in these same sectors in a new manner – as the maker and enforcer of the rules of the market. A legal infrastructure does not arise naturally from the dismantling of the planning system, however. This necessitates the construction of a new framework of regulation, or the modifications of present regulations, and a system enforcing these laws and regulations.

Rather than attempting to address all of the problems of China's intergovernmental relations and regulatory framework, this chapter focuses specifically on the issue of how China's existing intergovernmental power division has affected the functioning of the market system. Detailed discussions on intergovernmental relations in the areas of fiscal and monetary management can be found in the previous chapters. Legal and regulatory issues pertinent to natural monopoly (anti-trust law), consumer protection, and producer protection (trademark, patent, contract law, securities law, and so on) are discussed in this chapter only to the extent that they are relevant to our theme.

The rest of this chapter is organized as follows. The first section provides an overview of the division of regulatory powers between the central and local governments in the areas of investment project approval, industrial planning, pricing, marketing, international trade management, and so on. The second section examines the three main areas where the decentralized central–local relations have negatively impacted on the functioning of the market. The third section sketches a framework that can help limit the local governments' ability to restrict market competition. The three basic elements of this framework are an Interregional Commerce Clause and related laws that prohibit local governments' restrictions on resource flows, a Fair Trade Commission that ensures all regulations issued by government agencies are competition compatible, and reform of the government's administrative structure in the direction of reducing anti-competition government interventions.

DIVISION OF REGULATORY POWERS BETWEEN THE CENTRAL AND LOCAL GOVERNMENTS

In China, state-owned enterprises (SOEs) and collectively owned enterprises (COEs) account for approximately 89 per cent of the total industrial output.[1] Each SOE is administratively affiliated with a supervisory agency or its 'department in charge'. Most COEs are supervised by the local governments. For example, most township and village enterprises, the bulk of COEs, are under the control of township or village governments. Whether an enterprise's 'department in charge' is the central government or a local government largely determines who has the power to regulate the activities of the firm. If a firm's supervisory agency is a local government (usually represented by one of its industrial bureaus), it is referred to as being 'owned' by the local government, while if its supervisory agency is a line ministry under the State Council, it is referred to as being 'owned' by the central government. Over the past 40 years, the ownership (referring to who supervises the firm) of the SOEs has been decentralized and recentralized several times. Since the early 1980s, the main trend has been a decentralization of ownership from the central government to the local governments. By 1985, centrally-owned SOEs accounted for only 20 per cent of the total industrial output from all enterprises at and above the township level.

How do these ownership rights originate? The traditional rule is that investment confers ownership. The level of government that supplied the fixed capital for the enterprise obtains ownership rights over that enterprise. The customary rule is summed up in the expression 'shei jian, shei guan, shei yong' (who builds the enterprise, administers it, and has control over the output). Therefore, the enterprises owned by the central government are those established with central government funds or those taken over from private owners after 1949. Regional levels of government obtain ownership rights in enterprises mainly through investment or historical tradition.

Although the supervisory agencies of the enterprises do not have the legal property rights of the firms, they have several major powers that make them the *de facto* owners. First, they have the control over the dispersement of the these enterprises' assets. More specifically, they can determine whether to sell a firm or implement a merger with other firm(s). Second, under the fiscal revenue-sharing system introduced since 1980, the most important implication of ownership is that 'who owns the firm gets the taxes and revenue remittance from the firm'.[2] Third, to varying degrees, a supervisory agency is responsible for approving the firm's arrangement of the management personnel, production plan, investment project, procurement, price and wage setting, marketing, and so on. Under the contract responsibility system adopted in 1987–8 and the Regulations of Transforming Enterprise Mechanism promulgated by the State Council in 1993, most of these decision-making powers are supposed to be granted to the enterprises. In reality, however, many firms' supervisory agencies retained substantial control powers. This is particularly true in the inland areas where market-oriented reforms proceeded relatively slowly.

Enterprise activities regulated by agencies other than the firm's supervisory agency mainly involve investment. Investment projects with costs below 30 million yuan for technical innovation and 50 million yuan for basic construction are subject to the approval by local planning commissions; those with costs above the thresholds are subject to approval from the State Planning Commission. As for the source of funds, each level of government controls the spending of their budgetary funds and extra-budgetary funds and the local branches of specialized banks control the credit allocation between firms, given the credit ceiling assigned by the upper level bank branch. The allocations are influenced by each localities' industrial policy, which ranks the priority of each sector.

In the areas of production, material distribution, price and wage setting, the central government now retains only limited control over some key enterprises or products. In 1994, approximately 4.5 per cent of the total industrial output was produced according to a production plan formulated by the State Planning Commission; in 1993, only 13 major products (steel, coal, cement, and so on) were distributed by the Ministry of Materials and

Equipments under the coordination of the State Planning Commission, and by the end of 1992, the prices of approximately 5.9 per cent of total retail sales and the prices of 18.7 per cent of industrial raw materials in the country were controlled by the State Price Bureau.[3] The State Labor Department still stipulates that the total wage growth of all SOEs be limited to either productivity growth or the growth of gross profits. All local planning commissions, local bureaus of materials and equipment (abolished in mid-1993), price bureaus, and labor departments, impose similar controls over locally owned enterprises. Although the degree to which such controls are exercised varies greatly across regions, the general trend is that they tend to be less restrictive than the central government control over the centrally-owned SOEs. This is because controls over production, material allocation, prices, and wages are usually based on considerations of spill-over effects across sectors or across regions and concern the central government more than the local governments. According to many locally owned enterprises, 'Local governments are more interested in receiving revenues than controlling our operations.'[4]

In terms of international trade, the central government reserves its control over import and export rights granted to individual firms. After 10 years of decentralization, by the end of 1980s, there were approximately 4000 provincial and county level foreign trade corporations (FTCs) and several thousands of production enterprises which were granted the right of direct foreign trade without going through FTCs. All firms without direct export rights must be represented by a designated FTC in export business. Another mechanism through which the center controlled trade was the foreign exchange retention system, which stipulated that a certain percentage of foreign exchange earnings must be sold to the central government at the official exchange rate. Similar to the central–local fiscal relations, the Ministry of Foreign Economic Relations and Trade (MOFERT) used to sign contracts with each provincial level bureau of foreign economic relations and trade (FERT) on quotas of foreign exchange earnings. This foreign exchange retention policy was abolished in January 1994. The MOFERT also allocates import quotas for many goods subject to import control. Given the import quotas allocated by the MOFERT, the provincial FERT bureaus reallocate them to lower level FERT bureaus and firms. For importing goods subject to restrictions, the firms must apply to the FERT at the appropriate level (locally owned firms apply to the local FERT and centrally owned firms apply to the MOFERT).

As a major characteristic of the transitional period, there are numerous grey areas where the division of power between the central government or local governments is not clearly defined by any law or regulation. In these cases, the rights retained by the local governments are *de facto* and the center empirically recognizes the rights that the local governments claim. This often leads to power abuses by the local governments who take advantage of the lack of written laws or regulations. The center will then be forced to stipulate that certain rights do not belong to local governments and decisions in certain areas must be subject to approval of certain departments under the State Council. Two examples best illustrate such learning-by-doing experience. The first case is the cleansing of locally approved development zones. In 1992 and early 1993, there emerged nearly 1800 development zones approved by various levels of local government, including county, township, and village level government. Many local governments offered foreign-invested enterprises with tax exemptions and tax reductions that were more aggressive than those in the special economic zones approved by the State Council. Tax competition arising from the 'zone fever' had a serious impact on the government's revenue.[5] By mid-1993, the State Council announced that no local government below the provincial level had the authority to establish such zones offering tax concessions.[6]

Another case is about unauthorized local governments' transfers of 'land use rights'. During 1992–3, all levels of local governments started contracting land use rights for commercial use, as an attempt to attract external investment and increase local revenues. In most deals the land was undervalued and corruption and speculation spread widely. From January 1992 to July 1993, only 0.78 per cent of land transfer deals were made through bidding, 0.32 per cent through auction, while nearly 99 per cent of the deals were through free administrative transfer or contract transfer. In late 1993, the State Council stipulated that transfer of land use rights should in general be made through auction and bidding; in exceptional cases, contract transfer is permitted but the deal must be approved by the State Council or the government above the level conducting the transaction. The main elements of the State Council regulation on land use rights will be standardized in the 'Urban Real Estate Law', which is being reviewed by the National People's Congress (NPC).[7]

Recognizing the importance of a legal framework, the NPC has moved quickly in making new laws over the past few years. In 1993 alone, the NPC Standing Committee passed 20 laws and 13 provisional regulations, and deliberated over eight other draft laws, surpassing all recorded figures in its history. So far China has established a system of laws and regulations that mainly include various tax laws and regulations, the Anti-Unfair Competition Law, the Contract Law, the Patent Law, the Trademark Law, the Copyright Law, the Law of Foreign Investment, the Corporate Law, and regulations regarding consumer protection, securities trading, and futures markets. Laws regarding government budget, central banking, foreign trade, social security and insurance, foreign banks, and financial institutions are still under review or discussion. Nevertheless, it seems that inadequate attention has been paid to the serious impact of local governments' anti-competition behaviors on the functioning of the market. Accordingly, the legal system lacks an umbrella law that defines or helps define the limits of the local governments' regulatory powers. The next two sections will turn to these issues.

IMPACT OF DECENTRALIZATION ON THE MARKET SYSTEM

During the decentralization process over the past 15 years, the goal of promoting the function of the market in resource allocation has not been fulfilled satisfactorily. One major reason is that the many administrative powers that were supposed to be rendered to enterprises were in effect retained by different levels of local government, and were exercised, and often abused, with great discretion. This section discusses the negative impacts of decentralization on the functioning of the market system in three areas: interregional commodity trade has been restricted by local governments; interenterprises assets transfers (including mergers and acquisitions) are often opposed by the firms' supervisory agencies, most of which are local governments' industrial bureaus; and legal protection of interregional transactions is undermined by local government intervention. Central government policies adopted to address these issues are also examined.

Barriers to interregional trade

At the same time as the administrative decentralization, significant progress was made towards marketization in China. The important elements of this process were the relaxation of the central government's control over prices and the reduction of the central government's mandatory procurement. It was envisaged that the reduction of government intervention in commodities circulation would help form an efficient market system.

However, a free and unified market did not emerge automatically. Rather, the problem of regional market segmentation arose. Since 1985, as the economy became overheated, competition over raw material supplies intensified between regions. Using the administrative powers obtained from decentralization, local governments set up various forms of interregional trade barriers to stop the export of materials to other regions. These barriers were observed at both the provincial and the county levels and, consequently, the national market was segmented into little pieces. A large amount of anecdotal evidence was found in newspapers during 1985–8, among which the most famous were the 'cotton war', 'wool war', 'tobacco war', 'tea war', 'silk war', and 'grain war'. During these wars, prices of these commodities surged, a large proportion of production capacity in the coastal processing industries were wasted due to lack of materials, and exports and imports of raw materials increased simultaneously. When the economy went through a period of monetary contraction during 1989–90, trade barriers blocking finished products from entering local markets also emerged.

Typical forms of trade barriers set up by local governments included the following.

(1) Export embargo. In some areas, regional barriers took the form of the local authority organizing government officials or even local military force (Minbin) to blockade the transport which took the commodity out of their district. During the tea war, government cadres from Sheng county, Zhejiang Province were sent out along the county border to block peasants' trading with buyers from neighboring counties.[8] According to some accounts, it was these actions which led to the coining of the term 'commodity wars'.
(2) Transport license. When producers were tempted to sell their products to outsiders who offered market prices higher than the government procurement prices and the government procurement plans were hardly fulfilled, local authorities attempted to regain control over all purchase channels. For example, during the 'tea war', Zhejiang Provincial Government issued a notice stating that the government-controlled Market and Supply Cooperatives were the only designated purchasers of tea and all exports of tea to other provinces were subject to approval from the provincial government. To enforce these policies, both Zhejiang and Yunnan introduced transport licenses to exclude buyers from other provinces.
(3) Levy on exporting products. Some provinces imposed charges or fees on exporting products. Xinjiang, for example, placed a levy of 30 yuan per ton on exported raw wool.[9] This practice was also applied to commodity trade between counties and other levels of authority within a province.
(4) Request for rewards in exchange for 'exports'. During the cotton war, cotton-producing regions refused to supply raw cotton at the state planned price without receiving some other rewards, such as foreign exchange, low interest loans, or goods at a reduced price (for example, fertilizer at the planned price).[10]

(5) Strengthened enforcement of production quota. In some areas, local authorities set local production and procurement quotas on enterprises, with fines or budget cuts for those who fail to meet their targets. If these enterprises failed to meet the quotas, alternatives were to suspend the supplies of steel, cement, and timber, or withhold cadres' wages and even summon them for investigation.[11]

(6) Integration of material supply with processing industry. Local governments preferred to see rents created by price distortions (high market prices and low procurement prices) captured by local processing enterprises in their jurisdiction, rather than see them captured by outside producers. One option to recapture some of the rents was for the raw material makers to integrate forward into processing. During the wool war, new policies for local processing were introduced by the governments of the main wool-producing areas (Gansu, Qinghai, Xinjiang, and Inner Mongolia), which were anxious to expand their own wool-processing industries as a means of increasing their income. The policy became known as 'own production, own use, and own sale'. In Qinghai, Xinjiang, Gansu, and Inner Mongolia, new investments were made at both the provincial and county levels. In 1986, for example, Gansu added 16 384 spindles and Xinjiang built 25 textile factories and some 15 000 spindles. These increases meant that most of the local wool production would be needed to satisfy the local processing capacity. Inevitably, this led to increasing local reluctance to supply outsiders. Purchased wool was held in local stores and not sold to outsiders despite the storage costs involved. To promote the growth of small local factories, some local authorities offered tax relief advantages. Though producing low-quality products, these factories were able to offer lower product selling prices to compete in the new market.

There were two main reasons why local governments were interested in setting up trade barriers. The first motivation was to fulfill the central government procurement plan under the multi-tier price system. During the commodity wars, the central government retained some price control and mandatory procurement plans while secondary markets (markets outside the government procurement plan, including free markets) emerged. As a result of the partial marketing reform,[12] many agricultural and raw material products became subject to a range of prices, including the planned (base) price, negotiated prices (also for government procurement), and free market prices. Raw material producers, mostly peasants, were tempted to evade the price control by selling to outsiders who offered market prices, instead of fulfilling procurement plans. With different enterprises and trading units offering various inducements and incentives directly to the producers, the government procurement plans could no longer be guaranteed. Local authorities had to resort to administrative means, including various interregional trade barriers, to fulfill the mandatory procurement targets.

The second motivation for local governments to resort to trade barriers was to meet revenue quotas set by the fiscal contract with the center and/or to increase locally retained revenues. Under the fiscal contract system (see Chapter 1), local authorities were given the incentives and the administrative means to collect more revenues. Trade barriers, whenever appropriate for generating revenues or preventing outflows of local revenue, would be used. When the prices of agricultural and raw material products were controlled at artificially low levels, the local authorities' first reaction was to stop these products from 'exporting' to other regions. Inexpensive materials, once retained in the home area for producing final products, meant higher profitability of the local manufacturing industry. Even if the prices of some raw materials were free in the market, many local governments still attempted to block material outflow in order to protect its high-cost manufacturing sector. By limiting material outflow, the local price of raw materials was kept low and the less competitive local industries could survive. In most regions, industrial enterprises were the main source of government revenues while agriculture, which provided raw materials, was a less important source. In addition, the financial losses of agriculture were usually less visible to the local governments since most producers were poorly organized peasants.

The cost of market segmentation was substantial. The trade barriers increased the scarcity of the raw materials and boosted the prices outside the producing regions. As a result, the profitability of those technically more productive manufacturers outside the material-producing areas fell drastically. Between 1985 and the first half of 1988 the profit rate of four major wool textile factories in Beijing dropped from 19.2 per cent to only 0.2 per cent and the wool textile industry of the city as a whole was expected to make a loss. A similar prospect faced Tianjin while Shanghai was only avoiding losses because it still had stocks of 1987 wool.

Many of the most efficient and modern factories simply could not get supplies because of the competition from the large number of small enterprises established by local governments in material-producing regions. Production shifted from these efficient enterprises to local small factories where costs were higher, quality was poor, and pollution was not controlled. Moreover, the sharply increased local production capacity quickly exceeded the market demand as the market turned sluggish at the end of 1988. During the 1989–90 recession, an estimated 2.26 million spindles were operating with an annual capacity of around 220 000 tons against a domestic output level of around 50 000 tons.

Trade barriers also induced an inefficiently large volume of material imports. For example, the wool war led to a rapid increase in wool imports, particularly significant for the coastal processing industries. Imported wool, though until 1990 more expensive than the domestic product, was supplied in reliable and standard qualities and thus still enabled a good profit to be made. In addition, the state commonly subsidized the price of the imported wool to the processors so that the price differential was less significant than it might have been. There were also examples of one province (Hunan) restricting the supply of its raw material (silk) to another locality (Shanghai), preferring instead to export it abroad (Hong Kong) and the deficit locality (Shanghai) then had to import the raw material from abroad (Hong Kong).[13]

Since the early 1990s, regional trade barriers have been substantially reduced, particularly in the coastal areas. Several factors have contributed to this encouraging change. First, the prices of many agricultural products were gradually liberalized during this period and the state discontinued its mandatory grain procurement in most regions. For example, grain price in approximately 90 per cent of the counties was freed by the end of 1993. The price and marketing liberalization eliminated an important incentive for local governments to set up trade barriers (that is, they no longer need to meet the state procurement plan). Hunan Province, which originally was famous for restricting grain outflow to Guangdong, is now actively promoting the marketing of grain to other provinces. Second, in many coastal provinces, where government organizational reform had proceeded relatively quickly, some bureaus that had the power to erect trade barriers were streamlined or abolished. For example, in Guangdong, there is no longer a Bureau of Grain. Such reforms directly eliminated the instruments that many regions used to restrict interregional trade. Third, many regional leaders began to realize that the trade restriction may be harmful to their local economies. Some of them admitted that they made mistakes in the late 1980s by restricting raw materials from flowing out: they ended up with huge stocks of unmarketable materials when the market turned sluggish.[14]

However, there are continued accounts about trade barriers in many inland provinces. For example, a report said that until recently Sichuan's pork was not allowed to enter the market of Hunan and Hunan's cigarettes could not be sold in Hubei.[15] In Heilongjiang, some cities' and counties' Alcohol Monopoly Bureaus confiscated beer produced by Jixi City Brewery in order to reserve their markets for local producers.[16] In these inland areas, price liberalization was relatively slow and local governments were more eager to retain their power to intervene in firms' daily operation and were more reluctant to streamline their governmental functions. It is important to note that a number of factors that caused regional trade barriers still remain in place, although to a lesser extent compared with those in 1985–90. First, the nature of the current central–local fiscal sharing system still provides the localities with incentives to protect local enterprises that generate revenues. Second, the administrative powers retained by the local authorities, particularly in less developed regions, are still extensive. The local governments can easily intervene in enterprises' investment, production, and marketing. Third, while many prices of raw materials are being liberalized or adjusted closer to market prices, some are still under control and the plan–market price differential may again explode when demand surges. The co-existence of different marketing systems (that is, the state procurement system and the free market system) still implies continued competition for supplies.

Barriers to industrial reorganization

One of the major elements of the decentralization in the 1980s was the transfer of the *de facto* ownership of most SOEs from the ministries under the State Council to local governments. The *de facto* ownership of SOEs by local governments and their industrial bureaus dictates that new investments as well as industrial reorganization (including mergers and acquisitions) are largely confined to the administrative boundaries. This created the problem of duplication of investment (lack of specialization) and small scale of enterprises (lack of economies of scale). To address these issues, the central government adopted a number of measures to promote the formation of enterprise groups since the early 1980s. The stated objectives of this policy include promoting mergers and acquisitions among enterprises, reducing the regions', ministries', and industrial bureaus' monopoly powers, and increasing enterprises' management autonomy.

The origin of enterprise groups was the 'horizontal linkages' initiated in the early 1980s. In 1980, the State Council issued a circular stating that 'we should reorganize the industrial structure to avoid the problem of large number of small scale investments, duplication and blinded investments, according to the principle of specialization and interenterprise cooperation'.[17] It called for cross-region and cross-sector interenterprise cooperation in terms of technology transfer and assistance, investments, and long-term procurement contracts.

Based on the development of the 'horizontal linkages', enterprise groups emerged in 1986 and have gone through three stages. The first stage started in March 1986, when the State Council issued 'The Regulation of Selected

Issues Regarding Further Promoting Horizontal Linkages'.[18] At this stage, economic linkages between enterprises were largely confined to marketing cooperation between suppliers and purchasers, technology transfers from technologically advanced firms to other firms, contractual relations between assemblers and parts-suppliers, and so on. The second stage started in March 1987, marked by the announcement of 'The Interim Regulations of Independent Planning Status of Large Industrial Joint Enterprises'.[19] Since then, a number of enterprise groups were granted independent planning status. The third stage began with the 1988 document, 'Views on Developing Enterprise Groups', issued jointly by the Economic System Reform Commission and State Economic Commission.[20] A number of measures to promote mergers and acquisitions among enterprises, such as contracting, leasing, and corporatization have been experimented with since then.

Forming enterprise groups was aimed at breaking the administrative barriers and promoting resource reallocation among firms. However, it often faces opposition from the local governments and their industrial bureaus who have the *de facto* ownership of the enterprises. A typical case is as follows. There is a strong firm and a weak firm, subordinating to different local governments or different industrial bureaus under the same local government. The strong firm has marketable products and better technology, while the weak firm has abundant labor and/or equipment but lacks a market due to either substandard technology or poor management. These two firms are willing to merge, but the supervisory agency of the weak firm objects to such a merger.[21] Without appropriate compensation, the weak firm's supervisory agent will not allow its subordinate to be merged by firms outside its jurisdiction because it means a loss of revenue and ownership.

The central government policy adopted to circumvent this difficulty was named the 'three-no-change policy': if one firm wants to join an enterprise group, there should be no change in the firm's administrative supervisory agency, no change in the channel through which the firm remits its profits and taxes, and no change in the firm's ownership. This policy is a compromise of the central government's intention to break administrative barriers to industrial reorganization and the existing interests of the regions and industrial bureaus. This compromise has enabled the reform to proceed without encountering much opposition from the local governments. Without such a policy, forming enterprise groups across regions or industrial bureaus would have been impossible in many cases. The three-no-change policy made this transaction possible. To see this, suppose firm x in region X wants to merge will firm y in region Y while y still generates profits. Y is likely to reject the merger proposal since it means a loss of profit from y. Under the three-no-change policy, however, x can sign a management contract with Y, promising to submit a certain amount of profit to Y in exchange for acquiring the management rights and the residual profits (but not the ownership) of y.

Since firms' ownership is not transferable under the three-no-change policy, enterprise groups with tightly linked members through cross-shareholding are rare. Members of most groups are loosely-linked and the relations between members are often based on management contracts, procurement-supply contracts, and technical assistance. Financially very few groups are consolidated, and management and production plans of most member firms are independent. The impossibility of ownership transfer greatly restricts the degree to which the groups can reallocate internal resources and achieve the goal of greater specialization and economies of scale.

While the current enterprise group policy was aimed at promoting resource reallocation among firms, many measures adopted to achieve such a goal may be harmful to market competition for a number of reasons. First, currently many mergers and acquisitions are administratively initiated, rather than market-based ownership transfers. A typical form of such mergers is an industrial bureau instructs two firms both 'owned' by the bureau to merge. These administrative mergers are often opposed by the firms involved.[22] During the recent government organizational reform, many industrial bureaus that are to be abolished rename themselves as 'enterprise groups' and force their subordinate firms to join the groups. The motivation of these transformations was obviously to retain the powers that used to be in the hands of the industrial bureaus. Many of the newly established groups (transformed from old industrial bureaus) recentralized the management powers from the member firms, eliminated the member firms' legal person status, and imposed various arbitrary charges on them. It is recognized by the central government that such practices are monopolistic in nature and are detrimental to the development of the market.[23]

Second, many large and successful enterprise groups enjoy various special policies granted by the central and local governments. These policies include the following.

(1) Preferential profit retention and credit policy.[24] Many enterprise groups with preferential policies also enjoy low interest loans.
(2) Preferential procurement policies. The government offers many large enterprise groups procurement contracts and guarantees the supply of certain materials and energy products.
(3) Independent planning status granted to certain enterprise groups. The State Planning Commission granted 25 large enterprise groups such a

status in 1993. Many regional level enterprise groups are also given similar treatments by the local authorities.

These policies in effect discriminate against other firms or groups that are not on the government list. Furthermore, it cannot be justified by industrial policy since these policies are determined on a firm-by-firm basis.

Local protectionist intervention in legal cases

China has enacted many business-related laws in the 1980s, including the Contract Law, the Trademark Law, and the Patent Law. Recently, new laws and regulations on securities trading, anti-unfair competition, and copyright were promulgated; the Contract Law, the Trademark Law, and the Patent Law were revised to extend coverage. Speaking generally, the goal of these laws is to safeguard legal business transactions by defining the rights of buyers/sellers or producers/consumers and the penalty on behaviors that violate the laws. Without protection by laws, costs of business transactions would be prohibitively high and the transactions actually taking place would be inefficiently low.

While the promulgation of these laws is no doubt important, it is only a part of the legal system. Unfortunately, in China, respecting other party's legal rights in a business transaction has more of a sense of moral obligation than absolute rights – there is very little concept that the laws are binding. Over the years after many business laws were issued, they were poorly implemented. Practices that violate the Contract Law, the Patent Law, the Copyright Law, and the Trademark Law are common, and most of these violations are not settled through legal channels. Among various reasons, one of the most important up to now has been local protectionism.

In some cases, the local courts may unreasonably rule in favor of local parties. In 1991, the Ziyuan Textile Factory in Ziyuan county of Guangxi Province signed a sales contract with the Youfang Company of Hunan Province, a textile product retailer. Until 1993, Youfang still owed Ziyuan Textile 82 580 yuan. To settle this debt problem, the two parties reached an agreement according to which Youfang should pay Ziyuan 15 000 yuan and send 65 542 yuan worth of products to Ziyuan. When the check and the products arrived Ziyuan, the director of the Ziyuan Textile Factory declared that the agreement was invalid and then he confiscated Youfang's truck. The director of Ziyuan then appealed to the court of Ziyuan county for property protection for the truck. The county court ruled in favor of Ziyuan and imposed a penalty of 15 000 yuan on Youfang. Representatives of Youfang appealed to Guangxi Provincial Supreme Court, which later overruled the court judgement of Ziyuan. Under the pressure from the provincial court, the county court issued a ruling asking Ziyuan Textile to return the truck to Youfang. However, the county court did not enforce this ruling, letting Ziyuan Textile keep the truck. The chief judge of the county court said that he was not able to release the truck because the county leaders did not want to do so. He added, 'Now, power is bigger than law, and I can do nothing.' Since appealing to the court did not help, some workers of the Youfang Company seized a car from the Ziyuan county government in return. A simple business deal led to chaos.[25]

In other cases, the local judges refused to take action on cases against domestic parties. One major foreign bank lent a SOE the funds to help build a joint venture hotel in Tianjin. Months after the loans become overdue, the bank still had not been repaid. The bank wanted to recover the money, but the enterprise wanted concessions. The bank quickly learned that the outcome of the contract disputes is more a question of agreement than the application of principles. Intent on recovering its loan, the bank's lawyer approached the government. But the government and the state enterprise 'sit together on same side of the fence'. An officer of the bank said 'the attitude of the whole town was, it's our hotel and you are not going to take it way from us'. Going to court did not help either. Reluctant to execute an order that would enable the bank to assume control of the hotel, the presiding judge refused to take quick action. As the months dragged on, the bank eventually restructured the loan.[26]

In addition, a local court may refuse to cooperate with a court from another jurisdiction in executing the latter's decisions. A district court in Beijing Municipality ruled that a firm in Jiangsu Province violated the Trademark Law and caused financial losses to a firm located in Beijing. The Jiangsu firm was ordered to pay the Beijing firm compensation. According to the Trademark Law, the Beijing district court could request the local court in Jiangsu Province to enforce this ruling. However, the Jiangsu court simply refused to do so. The Beijing district court had to send its staff to Jiangsu to collect the penalty.[27]

The pressures of local interests on lower level courts is one of the chief factors detrimental to the Chinese judicial system. Offenders often ignore court judgments with complete impunity. Many people have informal channels to resolve disputes, even though a court may have already tried. Local power groups – business run by government officials or their associates – are particularly adept at protecting their interests. In many cases, this takes the form that local firms lobby a local government leader and the local leader expresses his/her 'opinion' on how to resolve the case to the local court. Sometimes lobbying translates into outright bribes and poorly paid judges make easy marks. Local court judges who rule in favor of the local parties are often considered, by the local leaders, as acting in the interest of 'our people'. Besides, local courts are subject to various subtle controls from the local governments: much of the local court staff's housing and benefits, as well as their children's education are provided by the local governments.

The problem is compounded by the fact that many judges are retired military officers with little or no legal training. The central and local governments have taken halting steps to address the problem, starting training programs for some judges. The more severe problem is the shortage of lawyers. Currently there are more than 5 million enterprises across the country, of which only 140 000 have legal advisors, accounting for 2.9 per cent of the total. In 1992, China's courts handled 690 000 economic cases, of which only 26.4 per cent were represented by lawyers.[28]

However, there are encouraging signs of change. One is the broad effort to boost the role of lawyers in China's legal system. The Ministry of Justice has mapped out its general target – by the end of 1995 and this century, the number of qualified lawyers should increase to 75 000 and 150 000 respectively, notary personnel should reach 20 000 and 30 000, respectively, and legal service people should increase to 130 000 and 200 000, respectively. To achieve this goal, the ministry plans to establish a complete system of examination, training, and professional technical standards for legal personnel, and to hold one lawyer qualification exam each year, replacing the every-other-year routine. Administratively, the ministry also calls for setting up non-government-funded law agencies. In some areas, government notary organizations will be transformed into non-governmental and non-profit organizations.

DEFINING THE LIMITS OF LOCAL GOVERNMENT POWER: THE RELEVANCE OF INTERNATIONAL EXPERIENCE

The previous section has argued that for the sake of protecting local interests local governments have the strong tendency to exercise/abuse their (vaguely defined) regulatory and administrative powers in a way that restricts the functioning of the market system. Realizing this, the central government has adopted a number of measures over the past years. However, many of these measures appear to be in the direction of administrative recentralization. For example, the industrial policy announced in 1989 partially aimed at addressing the issue of investment duplication and lack of economies of scale resulted from regional protectionism. It attempted to restrict or discourage investment in certain sectors that are favored by regional governments, such as small-scale textiles, cigarettes, and household electric appliances and at the same time encourage the sectors that are of 'national importance', such as energy, raw materials, and high tech industries.[29] While tax incentives and interest rate differentials were used as instruments to achieve these goals, an important measure was to recentralize the investment project approval rights for projects to be discouraged. Such recentralization policy brings with it the old problems in the planning system: complicated procedures, insensitivity to the consequences of decision making, and rent-seeking behavior.

Among the policies promoting enterprise grouping, one of the most important is that the center grants the large enterprise groups independent planning status. This measure raises the administrative ranks of the groups which may help them break the administrative barriers erected by governments and industrial bureaus at lower administrative levels. These types of policies, although conducive to breaking interregional barriers, may in effect foster administratively backed monopoly by high-rank enterprise groups. Often, these new groups with independent planning rights are at ministry or semi-ministry levels; their behavior will not be very much different from those of the line ministries.[30]

Rather than using administrative means, this section proposes a new framework to remedy the protectionist behavior of local governments. This framework involves laws that define the limits of local government regulatory powers and institutions to enforce these laws. The three basic elements of this proposal are the Interregional Commerce Clause and related laws that prohibit local governments' restrictions on resource flows across regions, the Fair Trade Commission that ensures all regulations issued by government agencies be competition compatible, and reform of the government administrative structure in the direction of reducing anti-competition government interventions. The next three subsections will discuss these questions in order.

Legislation on interregional commerce

The situation of interregional trade barriers in today's China is very much like that in the early years after the American Revolution. During the period of confederation (1781–7), protectionist state restrictions on interstate trade were common. Each state was free to pass tariff laws taxing goods from other states. New York, for instance, taxed firewood from New Jersey. To address this issue, the new constitution passed in the Philadelphia Convention granted Congress the power to regulate commerce between states. Under federal control, no state might levy tariffs on goods from other states.

Section 10 of Article 1, the Commerce Clause, of the constitution formally stated as follows:

> No state shall, without the consent of Congress, lay any imposts or duties on imports or exports, except what may be absolutely necessary for executing its inspection laws; and the net produce of all duties and imposts, laid by any states on imports or exports, shall be for the use of treasury of the United States; and such law should be subject to the revision and control of the Congress.

Congress's power to regulate commerce among the states has played an important role in the early years of the US economic development. Over time, this 'Commerce Clause' has come to have two main thrusts. First, in accordance with the original purpose, it is an independent check on state regulations that unduly restrict interstate commerce.[31] Today, the Commerce Clause is effectively an all-purpose federal police enabling Congress to regulate all sorts of activities within a state's borders (intrastate matters). Although the literal language of the Commerce Clause simply gives Congress the power to regulate commerce among the states, the Supreme Court has established the power of Congress to regulate intrastate matters by concluding that the power to regulate interstate commerce includes the ability to reach intrastate activities that have some impact on commerce between the states. In a highly interdependent society, it is difficult to find intrastate activities without some impact on interstate commerce. As a result, very few intrastate matters are outside the reach of the commerce power today.

In today's China, adding an Interregional Commerce Clause, which emulates the US Commerce Clause, into the constitution may be an important first step in constructing a legal framework that limits local governments' power abuses. The legislation should prohibit any government activities or regulations, at the central, provincial, county, and lower levels, that impede economic transactions between regions. Based on this clause, the law-makers should devise an Interregional Commerce Act, or provisions of the forthcoming Anti-Trust Law, to stipulate explicitly that any form of tariff or non-tariff barriers to interregional commodity flow, such as those listed previously, are unlawful and are subject to legal prosecution by a designated central government agency. In the short-term, the emphasis of promoting interregional commerce should be placed on commodity flow rather than on capital or labor mobility due to the enormous complications of the latter. The impediments to capital flow due to the central bank's region-based credit allocation cannot be removed before the reform of the monetary management system based on credit allocation. Labor mobility will be limited for an even longer period of time due to the various urban–rural policy differentials. In the long run, however, the objective of a common capital and labor market should also be pursued as a part of the legislative effort.

The Interregional Commerce Clause can serve as an umbrella legislation in defining local governments' administrative and legislative powers. Any regulation or administrative decision made by local governments that will be or potentially can be harmful to interregional commerce should be banned, even if such a decision is seemingly a 'domestic issue'. The clause will thus have a far reaching power in protecting the formation of a common market system.

Fair Trade Commission

Simply writing a clause in the constitution is only a first step to limiting local government powers. Institutionally, a central government agency must be set up to monitor the use of administrative powers at various levels of government, in particular at the local levels. This agency would operate under the legal framework centered on the Interstate Commerce Clause and other laws related to anti-trust/competition. At the same time this agency enforces competition-related laws, its important role should be to restrict the promulgation of and to remove the existing government policies that restrict competition or lead to unfair trade practices.

The experience of Korea in this regard provides some useful lessons. Over the past ten plus years, Korea's Fair Trade Commission lists 'reform of anti-competitive government regulations' as the fist item in its list of major activities. The commission reviews draft regulations to prevent the introduction of restrictive elements and every ministry which intends to issue regulations must consult the commission.[32] For example, there were 2640 applications for international agreements between 1981 and 1985, of which 931 or 35.3 per cent were judged by the Fair Trade Commission to contain provisions restricting competition or using unfair trade practices and were revised accordingly. There were 1850 applications for agreements on importing technology, the most common kind of agreement and 811, or 43.8 per cent, of these applications were ordered to make revisions.[33]

In China, the State Administration of Industry and Commerce (SAIC) is designated by the State Council to enforce the recently enacted Anti-unfair Competition Law[34] and other forthcoming competition-related laws such as the Anti-trust Law. However, past experience has shown that due to its low administrative rank – it is a semi-ministry level government organ – the SAIC has virtually no influence on any administrative decisions made and regulations promulgated by ministry-level government agencies, such as the line ministries under the State Council and the provincial-level governments. At the local level, the situation is similar: the power of local SAIC branches is very limited. Over the past 14 years, the main duty of the SAIC has been limited to the supervision of the distribution of goods by private proprietorships, the administration of the local consumer goods market for daily necessities, and enterprise registration.

It is hardly conceivable that the SAIC can be relied upon, with its present status, to undertake the task of preventing competition from being negatively impacted on by the anti-competitive policies of other government agencies. I suggest that a Fair Trade Commission, similar to that in Korea be created or transformed from the current SAIC. This new commission should be at the full-ministry level and be empowered by the Congress to review and approve all regulations enacted by government agencies at and below the ministry level. Charges against any government policy on the basis of violating the Interregional Commerce Clause and other competition laws will be handled by this commission. Examples of cases that the commission should handle include policies directly restricting interregional trade, local government protectionist procurement policies in favor of local producers and discriminating against outsider producers, licensing and technical standards discriminating against producers outside a government agency's jurisdiction, discriminative tax and financial policies not justified by the industrial policy, administrative intervention into court decisions related to interstate business, and government policies restricting enterprise mergers and acquisitions.

Needless to say, the coordination of economic policies between different government agencies and between different levels of the government is a very difficult task, in particular in China which is experiencing rapid changes in almost all aspects of the management system. However, a high priority should be given to protecting market competition and be supported by a visible effort of the decision makers throughout the reform process.

Government organizational reform

As indicated previously, the supervising agencies own SOEs and most COEs *de facto*. Such a relationship has enabled these agencies to adopt various administrative measures that restrict the functioning of the market. In the presence of these agencies, any reform attempting to render more management autonomy to the enterprises encounters resistance as the reform means less economic rents for government officials. The legal framework proposed above will also face serious challenge from these government agencies. Therefore, a government organizational reform to streamline and/or abolish some line ministries at the central level and the corresponding industrial bureaus at the local levels will be an essential step towards the establishment of a well-functioning market system.

China currently has a huge government with 40 million employees and this number swells by 1 million every year. In 1993, the State Council drafted the Government Organizational Reform Program, which was passed by the NPC. The program stipulates that the size of government staff will be cut by some 20 per cent in a year. The State Council will reduce the number of ministries and commissions from the current 86 to 59 and each level of local government should follow suit.

According to this program, reform of specialized ministries (in charge of specific industrial sectors) would take place within the following three categories.

(1) The Ministry of Aeronautics and Astronautics would be transformed into the China Aerospace Industrial Corporation and the China Aviation Industry Corporation. These corporations would be purely business entities with no administrative function.
(2) The Ministry of Light Industry and the Ministry of Textile would become industrial associations under the State Council. These associations and their local counterparts (former local light industrial bureaus and textile bureaus) would no longer control enterprise production planning, pricing, and investment project approval. The main functions of the two associations would be to improve trade planning, implement trade policies, and provide enterprises with information services.
(3) Other former specialized ministries would carry over or merge into newly created ministries. Local counterparts of all these reformed ministries would make similar changes according to the new structure at the central level. The new ministries and local bureaus would feature streamlined administrative structures and management functions.

The direction of the government organizational reform is undoubtedly consistent with our proposal in this chapter. However, the coverage of the reform should be extended to many other ministries and their local counterparts. To ensure a better functioning of the market, most specialized ministries and local industrial bureaus managing specific industrial sectors should be streamlined in the short-term and abolished in the long-term. In addition, during the organizational reform, special restrictions should be imposed on transformations of industrial bureaus into corporations with monopoly powers. For example, provisions of government organizational transformation should stipulate that new companies transformed from industrial bureaus cannot claim ownership of the enterprises that were formerly under these bureau.

CONCLUSIONS

This chapter looks at the impact of decentralization on the functioning of China's market system. It is argued that as the Chinese central government relaxed its control over the economy through decentralization, local governments turned many of the transferred powers to restrict market competition. Reasons for such local government behaviors include fulfilling fiscal contracts or increasing local revenue retention, meeting central government procurement quotas under a dual-track price system, and protecting local industries and employment. The *de facto* ownership of local enterprises by local governments enables the latter to realize these objectives through protectionist measures.

A legal framework designed to restrict local governments' power abuse is therefore necessitated. Currently, the division of regulatory power between the central and local governments is poorly defined: there are many grey areas where no guidelines restricting local government power exist. This chapter proposes that an umbrella legislation of Interregional Commerce be enacted and a Fair Trade Commission be created to ensure that government policies at all levels be pro-competition. As an institutional prerequisite, the government organizational reform should be extended to more sectors of the economy.

Appendix Tables

TABLE A.1 *Administrative division in China, 1993*

	Number of prefectures and prefectural level cities	*Number of counties and county level cities*	*Districts under city administrations*
All China	35	2 166	669
Beijing		8	10
Tianjin		5	13
Hebei	12	139	34
Shanxi	11	100	18
Inner Mongolia	12	84	16
Liaoning	14	44	56
Jilin	9	40	19
Heilongjiang	14	68	64
Shanghai		6	14
Jiangsu	11	64	42
Zhejiang	11	66	20
Anhui	16	68	35
Fujian	9	64	17
Jiangxi	11	84	15
Shandong	17	98	37
Henan	17	118	39
Hubei	14	70	29
Hunan	14	96	26
Guangdong	20	79	40
Guangxi	14	82	23
Hainan	2	17	3
Sichuan	23	178	43
Guizhou	9	80	6
Yunnan	17	123	4
Tibet	7	77	1
Shaanxi	10	93	14
Gansu	14	75	10
Qinghai	8	39	4
Ningxia	4	18	6
Xinjiang	15	85	11

SOURCE State Statistical Bureau (1994).

TABLE A.2 *China: basic indices by region, 1993*

	Population (million)	*GDP (billion yuan)*	*Per Capita GDP (thousand yuan)*	*Net industrial output (billion yuan)*	*Net agricultural output (billion yuan)*
All China	1 185.17	3 210.00	2.71	981.65	579.34
Beijing	11.12	86.35	7.77	21.34	4.23
Tianjin	9.28	53.61	5.78	20.52	3.14
Hebei	63.34	156.66	2.47	51.35	25.61
Shanxi	30.12	64.58	2.14	22.78	8.30
Inner Mongolia	22.32	48.66	2.18	11.16	12.14
Liaoning	40.42	180.82	4.47	65.17	20.11
Jilin	25.55	67.20	2.63	22.47	13.23
Heilongjiang	36.40	107.69	2.96	39.06	18.11
Shanghai	13.49	151.16	11.21	62.37	3.60
Jiangsu	69.67	275.45	3.95	102.50	39.57
Zhejiang	42.66	169.80	3.98	59.49	25.67
Anhui	58.97	0.96	1.66	30.17	24.41
Fujian	31.50	102.81	3.26	24.82	19.74
Jiangxi	39.66	70.19	1.77	17.64	20.19
Shandong	86.42	270.25	3.13	88.13	52.34
Henan	89.49	158.31	1.77	46.24	35.03
Hubei	56.53	129.84	2.30	37.90	30.27
Hunan	63.11	119.24	1.89	28.48	32.40
Guangdong	66.07	325.53	4.93	87.87	46.43
Guangxi	44.38	78.80	1.78	16.87	23.25
Hainan	7.01	22.52	3.21	2.01	5.77
Sichuan	111.04	195.87	1.76	50.07	48.85
Guizhou	34.09	40.85	1.20	10.20	11.80
Yunnan	38.85	66.22	1.70	18.51	18.25
Tibet	2.32	3.73	1.61	0.22	1.78
Shaanxi	34.43	61.45	1.78	18.71	12.69
Gansu	23.45	35.83	1.53	11.08	7.45
Qinghai	4.67	10.57	2.26	2.30	1.92
Ningxia	4.95	9.84	1.99	2.64	1.81
Xinjiang	46.05	48.19	1.05	9.58	11.26

SOURCE State Statistical Bureau (1994).

TABLE A.3 *China: sources of government revenue, 1980–93 (in billions of yuan)*

	1980	*1985*	*1990*	*1991*	*1992*	*1993*
Total revenue	108.52	186.64	331.26	361.09	415.31	508.82
1. Taxes	57.17	204.08	282.19	299.02	312.16	425.53
Industrial and commercial taxes*	50.14	114.78	189.54	202.47	227.73	331.95
Special tax on oil and fuel		1.70	1.14	0.99	0.73	0.16
Salt tax	0.92	1.00	0.84	0.84	0.81	0.78
Customs duties	3.35	20.52	15.90	18.73	21.28	25.65
Agricultural and animal husbandry tax	2.77	4.21	8.79	9.07	9.00	9.64
Construction tax		2.29	3.79	3.09	3.23	3.84
2. Revenue remittance from enterprises	43.52	4.38	7.83	7.47	6.00	4.95
3. Borrowings†	4.30	8.99	37.55	46.14	66.97	69.02
4. Funds for energy and transportation key projects		1.47	1.85	1.88	1.57	1.18
5. Other revenues	3.53	5.22	43.07	40.66	58.98	38.67
6. Subsidies to loss-making enterprises			(57.89)	(51.02)	(44.50)	(41.13)

* Mainly includes enterprise income tax, product tax, value added tax, and business tax.

† Includes revenues from domestic and foreign debts.

SOURCE State Statistical Bureau (1994).

TABLE A.4 *China: government expenditure by function, 1980–93 (in billions of yuan)*

	1980	*1985*	*1990*	*1991*	*1992*	*1993*
Total expenditure	121.27	184.48	345.22	381.36	438.97	528.74
Capital construction	41.94	58.38	72.56	73.98	79.30	90.08
Circulating funds	3.62	1.43	1.09	1.31	6.65	1.85
Technological transformation and three technology expenditures	8.05	10.34	15.39	18.08	37.79	42.14
Geological prospecting	2.26	2.96	3.62	3.83	46.34	4.91
Administrative expenses in industry, transportation, and commerce	2.29	3.52	4.69	5.24	43.86	7.62
Agricultural production and administration	8.21	10.10	22.18	24.36	32.16	32.34
Culture, education, science, and health care	14.63	31.67	61.73	70.80	79.30	95.78
Social relief and welfare	2.03	3.11	5.50	6.73	6.65	7.53
National defense	19.38	19.15	29.03	33.03	37.79	42.58
Government administration	6.68	14.36	33.35	37.58	46.34	58.58
Repayment of debts	2.86	3.96	19.04	24.68	43.86	33.62
Price subsidies			38.08	37.38	32.16	29.93

SOURCE State Statistical Bureau (1994).

TABLE A.5 *China: division of central and local government expenditure by function, 1990*

Items	*Expenditure (billion yuan)*	*As % of total expenditure*	*Central (%)*	*Local (%)*
Capital construction	72.56	21.02	75.66	24.34
Circulating funds	1.09	0.32	72.73	27.27
Enterprise technology transformation	7.80	2.26	8.52	91.48
Three technology expenditures	9.80	2.84	73.91	26.09
Geological prospecting	3.62	1.05	100.00	0.00
Administrative expenses in industry and transportation	4.21	1.22	31.90	68.10
Administrative expenses in commerce	0.48	0.14	11.49	88.51
Agricultural production	12.81	3.71	7.37	92.63
Agricultural administration	9.39	2.72	12.63	87.37
Culture, education, and health care	57.27	16.59	8.12	91.88
Science	4.45	1.29	56.68	43.32
Social relief and welfare	5.50	1.59	0.00	100.00
National defense	29.03	8.41	100.00	0.00
Government administration	30.31	8.78	8.75	91.25
Armed police	3.04	0.88	100.00	0.00
Repayment of debts	19.04	5.52	100.00	0.00
Price subsidies	38.08	11.03	10.50	89.50
Others	36.73	10.64	12.63	87.37

SOURCE Ministry of Finance (1992) and State Statistical Bureau (1994).

TABLE A.6 *China: extra-budgetary revenue and expenditure of the central and local governments (in billions of yuan)*

Year	*Extra-budgetary revenue*			*Extra-budgetary expenditure*		
	Total	*Central*	*Local*	*Total*	*Central*	*Local*
1982	80.27	27.07	53.20	73.45	22.71	50.75
1983	96.77	35.99	60.78	87.58	30.04	57.54
1984	118.85	47.05	71.79	111.47	42.02	69.45
1985	153.00	63.61	89.39	137.50	56.21	81.30
1986	173.73	71.66	102.07	157.84	64.09	93.74
1987	202.88	82.80	120.08	184.08	74.16	109.91
1988	227.00	90.72	145.36	214.53	84.29	130.24
1989	265.88	107.23	158.66	250.31	97.59	152.72
1990	270.96	107.33	163.54	270.71	103.77	166.94
1991	324.33	138.11	186.22	309.23	126.33	182.90
1992	385.49	170.77	214.72	364.99	159.28	205.71

SOURCE State Statistical Bureau (1994).

TABLE A.7 *China and selected countries: central and subnational government revenue, 1990*

Country	*Revenue*			*Revenue/ GDP ratio (%)*	*Share in total revenue*	
	Central	*Subnational*	*Total*		*Central (%)*	*Subnational (%)*
Australia (billion dollars)	100.4	39.1	139.5	36.9	72.0	28.0
Brazil (thousand reasi)	3628.2	1208.4	4836.6	14.8	75.0	25.0
China (billion yuan)	136.8	194.5	331.3	18.7	41.3	58.7
Canada (billion dollars)	138.8	151.7	290.4	45.1	47.8	52.2
France (billion francs)	2638.6	380.7	3019.3	46.4	87.4	12.6
West Germany (billion DM)	699.0	379.2	1078.2	44.5	64.8	35.2
Hungary (billion forint)	1105.9	136.6	1242.5	59.7	89.0	11.0
India (billion rupees)	723.5	351.4	1074.9	20.2	67.3	32.7
Indonesia (trillion rupiah)	39.6	5.0	44.6	22.8	88.8	11.2
Japan (trillion yen)	62.2	31.8	93.9	22.1	66.2	33.8
Romania (billion lei)	297.9	43.3	341.2	40.4	87.3	12.7
UK (billion pounds)	203.0	25.8	228.8	41.7	88.7	11.3
US (billion US dollars)	1085.9	775.1	1861.0	34.1	58.3	41.7

NOTE: central and subnational revenue figures exclude transfers from other levels of government.
SOURCE Calculated from data in IMF (1994) and State Statistical Bureau (1994).

TABLE A.8 *China and selected countries: central and subnational government expenditure, 1990*

Country	*Expenditure*			*Expenditure/ GDP ratio (%)*	*Share in total expenditure*	
	Central	*Subnational*	*Total*		*Central (%)*	*Subnational (%)*
Australia (billion dollars)	67.0	66.0	132.9	35.1	50.4	49.6
Brazil (thousand reasi)	3448.7	1883.7	5332.4	16.3	64.7	35.3
China (billion yuan)	137.3	207.9	345.2	19.5	39.8	60.2
Canada (billion dollars)	132.5	188.1	320.6	49.7	41.3	58.7
France (billion francs)	2554.9	581.2	3136.1	48.2	81.5	18.5
West Germany (billion DM)	656.6	432.8	1089.4	44.9	60.3	39.7
Hungary (billion forint)	987.2	218.6	1205.8	58.0	81.9	18.1
India (billion rupees)	643.0	672.8	1315.8	24.8	48.9	51.1
Indonesia (trillion rupiah)	31.5	4.9	36.4	18.6	86.6	13.4
Japan (trillion yen)	31.4	71.7	103.0	24.3	30.5	69.5
Romania (billion lei)	281.1	51.4	332.5	39.4	84.6	15.4
UK (billion pounds)	167.4	68.4	235.7	42.9	71.0	29.0
US (billion US dollars)	1182.1	848.7	2030.9	37.2	58.2	41.8

NOTE: central and subnational expenditure figures include transfers to other levels of government.
SOURCE Calculated from data in IMF (1994) and State Statistical Bureau (1994).

TABLE A.9 *China: number of financial institutions in the state banking system, end of 1993*

	Total	*PBC*	*ICBC*	*ABC*	*BOC*	*PCBC*	*PICC*	*BOCOM*
Institutions	143 796	2534	34 002	58 964	9886	32 456	4959	995
National headquarters	7	1	1	1	1	1	1	1
Branches at provincial level	191	30	29	29	30	30	30	13
Branches at direct planning cities	85	14	14	14	14	14	15	
Branches in prefectures and prefectural level cities	2115	315	422	312	363	327	314	62
Branches in counties	12 130	2064	2336	2226	1217	2130	2157	
City offices	7180	17	1134	635	1147	3172	801	274
Business offices	469			370				99
Small and local branches	41 505		3987	31 770	1759	2261	1512	216
Savings deposit offices	71 242		21 831	22 501	5238	21 388		284
Other	8872	93	4248	1106	117	3133	129	46

NOTES: PBC, People's Bank of China; ICBC, Industrial and Commercial Bank of China; ABC, Agricultural Bank of China; BOC, Bank of China; PCBC, People's Construction Bank of China; PICC, People's Insurance Company of China; BOCOM, Bank of Communications.

SOURCE State Statistical Bureau (1994).

TABLE A.10 *Indices of China's state banking system, 1978–94 (in billions of yuan)*

Year	*Total loans*	*Total deposits*	*Total assets*
1978	185.00	113.45	187.65
1979	203.96	133.91	216.26
1980	241.43	166.12	262.43
1981	286.02	202.74	317.07
1982	318.06	236.99	361.84
1983	358.99	278.86	412.49
1984	476.61	358.39	537.03
1985	590.56	426.49	637.45
1986	759.08	535.47	811.14
1987	903.25	651.70	987.04
1988	1055.13	742.56	1154.13
1989	1240.93	901.39	1361.79
1990	1516.64	1164.48	1683.79
1991	1804.40	1486.41	2061.39
1992	2161.55	1889.11	2426.91
1993	2646.11	2323.03	3419.46
1994	3160.91	3227.89	4582.26

NOTES: figures after 1980 include the Bank of Construction. Figures after 1987 include the Bank of Communications and CITIC.
SOURCE State Statistical Bureau (1989), State Statistical Bureau (1994), and Editorial Department of China Finance (1995).

TABLE A.11 *China: selected macroeconomic indicators, 1978–94 (in billions of yuan)*

Year	*GNP (current price)*	*GNP index (1978=100)*	*Retail price index (1978=100)*	*Currency in circulation*	*M1*	*Quasi-money*	*M2*	*Lending to MOF*
1978	358.8	100.0	100.0		58.0	30.9	89.0	
1979	399.8	107.6	102.0		92.1	40.6	132.8	
1980	447.0	116.0	108.1		114.9	52.2	167.1	
1981	477.3	121.2	110.7		134.5	63.3	197.8	
1982	519.3	131.8	112.8		148.8	77.7	226.6	
1983	580.9	145.5	114.5		174.9	96.4	271.3	
1984	696.2	166.9	117.7		244.9	114.9	359.9	
1985	855.8	188.2	128.1	98.8	301.7	185.8	487.5	27.5
1986	969.6	203.5	135.8	121.8	385.9	249.0	634.9	37.0
1987	1130.1	225.7	145.7	145.5	457.4	338.3	795.7	51.5
1988	1406.8	251.2	172.6	213.3	548.7	411.5	960.2	57.7
1989	1599.3	262.1	203.3	234.2	583.4	555.9	1139.3	68.5
1990	1769.5	272.7	207.6	264.4	701.0	767.2	1468.2	80.1
1991	2023.6	295.0	213.6	317.8	898.8	961.1	1859.9	106.8
1992	2437.8	334.5	225.2	433.6	1171.4	1261.3	2432.7	124.1
1993	3134.2	378.7	254.9	586.5	1424.4	1583.2	3007.6	158.2
1994	4491.8	423.4	310.2	522.0	2055.6	2637.7	4693.3	168.7

NOTES: M1 (money) equals the sum of currency outside banks and demand deposits other than those of the central government. Quasi-money comprises time, savings, and foreign currency deposits of resident sectors other than the central government. M2 is the sum of M1 and quasi-money.
SOURCE IMF (1994), State Statistical Bureau (1994), State Statistical Bureau (1995), and Editorial Department of China Finance (1995).

End Notes

NOTES ON THE INTRODUCTION

1. Taiwan Province of China is not included.
2. Figures at the end of 1993. See State Statistical Bureau (1994).

NOTES ON CHAPTER 1: INTERGOVERNMENTAL FISCAL RELATIONS AND MACROECONOMIC MANAGEMENT

1. The following literature has described in detail the evolution of China's central provincial fiscal relations since 1980: World Bank (1990a, 1993, 1994a), Oksenberg and Tong (1991), Wong (1991), Agarwala (1992), and Bahl and Wallich (1992).
2. The central–local fiscal system adopted in 1988 was named The Big (comprehensive) Fiscal Contract System. In the rest of this paper, we broadly refer to all variants of the decentralized fiscal system adopted since 1980 as the 'fiscal contract system'.
3. The tax assignment structure (three types of revenues) announced in 1980 by the State Council was as follows (State Council, 1981): central-fixed revenues – profit remittance from centrally owned state enterprises, custom duties, and other income; local-fixed revenues – profit remittance from locally owned enterprises, salt tax, agricultural and animal husbandry tax, consolidated industrial and income tax, and other local taxes and income; revenues shared by the center and localities – income of local enterprises that have been placed under the direct administration of the State Council.
4. In 1985, the tax assignment structure based on three types of revenues was extended to all provinces except Guangdong and Fujian, and the items included in each revenue type expanded as a result of the 1983 tax reform. The tax assignment structure announced in 1985 was as follows (State Council, 1986).

   Central-fixed revenues: income tax, adjustment tax, and profit remittance of centrally owned state enterprises; business tax from railroads, banks and insurance company headquarters; income from enterprises in the defense industry; income from enterprises that signed fiscal contracts with the central government; fuel oil special taxes; all customs duties, product tax, and value added tax collected at customs; special adjustment tax; consolidated industrial and commerce tax and income tax on offshore oil activities of foreign companies and joint ventures; treasure bond income; state fund on energy and transportation key projects; 70 per cent of the three turnover taxes (product tax, business tax, and value added tax) collected from enterprises owned by the Ministry of Petroleum Industry, the Ministry of Power, China Petrochemical Corporation, and China Nonferrous Metal Company; and other income.

Local-fixed revenues: income tax, adjustment tax, and contracted revenue remittances from locally owned state enterprises; income tax from collectively owned enterprises; agricultural and animal husbandry tax; vehicle and ship registration tax; urban real estate tax; slaughter tax; domestic animal trading tax; market tax; title trade tax; income from penalties on delayed tax payments; urban maintenance and construction tax; land ultilization tax; vehicle utilization tax; 30 per cent of the turnover taxes (product tax, business tax, and value added tax) collected from enterprises owned by the Ministry of Petroleum Industry, the Ministry of Power, China Petrochemical Corporation, and China Nonferrous Metal Company; and other income.

Revenues shared by the center and localities: product tax, business tax, and value added tax (excluding those from enterprises owned by the Ministry of Petroleum Industry, the Ministry of Power, China Petrochemical Corporation, and China Nonferrous Metal Company, and turnover taxes from railroads, banks and insurance company headquarters); natural resource taxes; construction tax; salt tax; individual income tax; bonus tax on state-owned enterprises; and consolidated industrial and commercial tax, income tax levied on foreign and joint venture enterprises (excluding those on offshore oil activities of foreign companies and joint ventures). The 1988–93 tax assignment structure was largely based on the 1985 structure.

5. See Agarwala (1992) for details.
6. These methods were applied to 28 provinces, municipalities directly under the State Council, autonomous regions, and nine cities with independent planning status. The difference between the expenditures and income of Hubei and Sichuan provinces was made up from income that Wuhan and Chongqing cities sent to the provinces.
7. Using the standard definition. See Tables 1.1 and 1.2.
8. A referee says that 'the loss in revenue [declining revenue/GNP ratio] was most likely deliberate [a result of the center's deliberate effort]'. This was true to some extent in the early stages of the fiscal decentralization. However, since 1988, the revenue-sharing methods were largely designed for the purpose of increasing local revenue collection. The central government has repeated expressed concerns about the declining trend of the revenue : GNP ratio and almost all the efforts made in the last few years to reform the central–local fiscal relations were aimed at correcting this trend.
9. See 'Vice Finance Minister Xiang Huicheng Speaks on Reform in Tax-Assignment System', *People's Daily* (Overseas Edition), 23 November 1993.
10. 'Provisional Regulations Concerning the Introduction of a Fiscal Management System by Which the Scope of Revenues and Expenditures is Defined for the Central and Local Governments, Respectively, and Each is Responsible for the Balancing of Its Own Budget,' the State Council, 1 February 1980. See Economic Research Center of the State Council (1981) and World Bank (1993).
11. See Agarwala (1992).
12. Discussion of the rationale for these propositions can be found in Musgrave (1965), Oates (1972), Shah (1994), Agarwala (1992), and Persson and Tabellini (1990), among others.
13. Among the most influential studies is Hu and Wang (1994). Also see Wang (1994a).
14. See Development Research Center of the State Council (1993).

15. The following four paragraphs draw heavily from Ma (1993a).
16. Interview with officials of the State Council's SEZ Office.
17. See Ma and Bell (1992).
18. State Council's SEZ Office.
19. Interview with officials of the Ministry of Finance.
20. According to the state's industrial policy, enterprises in priority sectors (for example, electronics and export sector) meeting certain requirements can enjoy income tax holidays. For example, enterprises that export more than 70 per cent of their output value can enjoy income tax reduction. For newly established collectively owned enterprises, tax exemption applies for the first 2 years. Enterprises established by educational units, certain mass organizations, and social welfare units can also exempt or reduce income tax. See An *et al.* (1991).
21. See An *et al.* (1991).
22. See State Statistical Bureau (1993b).
23. Defined as (taxes + fees)/(net income).
24. See World Bank (1994b).
25. Evidence of central government borrowing from local governments, the central government's reclaiming ownership of local firms, and its transferring expenditure responsibilities to local governments is summarized in Yuan (1992). Examples of changes in fiscal relationships between the center and Shanghai, and between the center and Guangdong can be found in Jia and Lin (1994).
26. The picture is different if the center can commit to a pre-announced revenue-sharing schedule. Given the center's commitment, the regions will not react strategically and revenue collection will be higher than in the no commitment case. The center and the regions will both be better off under the commitment regime. See Chapter 2.
27. See World Bank (1993).
28. Looking at the relationship between the deficit : GNP ratio and the GNP growth rate has been a standard exercise for analyzing the role of fiscal policy in macroeconomic stabilization. This methodology can be found in several textbooks including Gordon (1990) and Prachowny (1994). A referee has correctly pointed out that, ideally, quasi-fiscal lending should be taken into account when measuring the actual fiscal deficit. But simply taking the sum of the budgetary deficit and total policy lending from the central bank as the consolidated fiscal deficit is not appropriate. A reasonable measure of consolidated fiscal deficit should be the sum of fiscal deficit and the implied subsidies of the policy lending, but the data of implied subsidies do not seem to exist.
29. The result of the ordinary least square (OLS) regression is as follows:

$$\text{DGR} = \underset{(2.698)}{9.586} - \underset{(0.150)}{0.090}\ \text{RGGR}$$

$R^2 = 0.001$, number of observations = 15, and degrees of freedom = 13. In the regression, DGR stands for the deficit : GNP ratio and RGGR stands for real GNP growth rate. Numbers in parentheses are *t* ratios. Data used are from 1979 to 1993. Data source: State Statistical Bureau (1992a, 1993b, 1994).

30. People's Bank of China (1993).
31. Using data from 1979 to 1993, a simple regression of revenue growth rate on nominal GNP growth rate shows only mild correlation between the two variables. The OLS regression result is as follows:

$$\text{RG} = 3.054 + 0.341\ \text{NGGR}$$
$$(0.427)\quad (1.244)$$

$R^2 = 0.106$, number of observations = 15, and degrees of freedom = 13. In the regression, RG stands for the growth rate of revenue, using the standard definition (See Table 1.1) and NGGR stands for nominal GNP growth rate. Numbers in parentheses are ratios. Date source: State Statistical Bureau (1992a, 1993b, 1994).
32. Revenue figures are those before tax sharing.
33. See 'Vice Finance Minister Xiang Huicheng Speaks on Reform in Tax-Assignment System', *People's Daily* (Overseas Edition), 23 November 1993.
34. However, this reduction will not significantly change the effective income tax rate for SOEs. According to one estimate, SOEs' effective income tax rate (as implemented in the contract system) was approximately 32.5 per cent in 1992 (interview with staff of the Ministry of Finance).
35. The method to determine local government revenues since 1994 is as follows. First, a local government would continue to remit/receive the contracted remittance/contracted subsidy to/from the center as stipulated in the 1993 contract. Secondly, the center would return the 1993 base figure (defined as consumption tax equivalent raised in 1993 plus 75 percent of VAT equivalent raised in 1993, adjusted for a number of revenue sharing items in 1993) to the local government. Thirdly, after 1994, the amount returned from the center to a local government is to increase from the 1993 base figure by 0.3 per cent for each percentage increase in the national total of VAT and consumption tax (this formula was later changed into one relating the returned revenue to the total of a locality's VAT and consumption tax). For a detailed description on how to calculate local revenues since 1994, see Zhang (1995).
36. Presentation by Yao Xitang, President of Pudong (Shanghai) Development Academy, on China's central–local relations at Washington Center for China Study (Washington, DC, August 1993).
37. See Appendix 1.1.
38. This appendix draws from Ma (1994a).

NOTES ON CHAPER 2: CENTRAL GOVERNMENT CREDIBILITY AND LOCAL TAX EFFORTS: A GAME THEORETIC MODEL

1. For a detailed description of the three systems adopted since 1980, see Chapter 1.
2. Both growth rates are calculated based on figures deflated by a GNP deflator. Data source: State Statistical Bureau (1994).
3. In recent years, the central government has repeatedly issued circulars banning unauthorized tax concessions and exemptions by local governments. See Chapter 1.

4. Major studies on China's central–local fiscal relations include, among others, World Bank (1990a, 1993), Oksenberg and Tong (1991), Wong (1991), Agarwala (1992), Bahl and Wallich (1992), Blejer (1993), and Jia and Lin (1994).
5. Although this model is motivated by the upward revenue sharing system (in which the regions collect most revenues and then share with the center) in China before 1994, it is also applicable to the case of a downward revenue sharing system (in which the center collects more than it spends and shares a part of the centrally collected revenue with the regions) as observed in many federalist countries. China's 1994 tax assignment reform is changing the system into one of downward revenue-sharing. To analyze the case of downward revenue sharing, one can assume that x_1 and x_2 are negative and represent central government transfers to regions 1 and 2 as percentages of the regions' own revenue collections.
6. Looking at the special case of $E = 0$ is an easy way to show that the center's redistribution motive is the source of the time-inconsistency problem discussed in the following sections.
7. To see this, suppose $a_1 < a_2$. Equations (2.18) and (2.19) state that $Y_i = (1 - x_i)/2a_i$, $i = 1, 2$. For equal tax rate x, region 1 collects more revenue than region 2:

$$Y_1 = (1 - x_1)/2a_1 > Y_2 = (1 - x_2)/2a_2.$$

8. See Fujiwara (1992).
9. This can be viewed as an application of the Folk theorem (see Fudenberg and Tirole, 1991).
10. Note that this statement is not based on the comparison of the Chinese τ with the critical value of 1/2 derived in the above model with arbitrary parameters.
11. In his comments on my paper, Prof. Roy Bahl offered the following explanation of why the center is unwilling to give up the flexibility in redefining central-local fiscal relations: the fear of inflation, the shortage of capital finance, and the desire to redirect investment flows has led the center to play the revenue sharing game by siphoning off additional resources from the locals, even at the expense of less revenue mobilization.

NOTES ON CHAPTER 3: CENTRAL GOVERNMENT TRANSFER UNDER A SOFT BUDGET CONSTRAINT

1. Between 1983 and 1991, the actual expenditures exceeded the budgeted expenditures by 7 per cent on the average. See World Bank (1993, p. 12).
2. In this chapter, 'transfers to localities' do not include 'contracted transfers to provinces' as defined in the 1988 fiscal contract system (pp. 7); contracted transfers are viewed as the localities' own revenues after tax sharing.
3. Interviews with officials of the Ministry of Finance, State Planning Commissions, and Development Research Center of the State Council.
4. This specification comes from Chapter 2. The conclusions do not change if we

assume that the center also takes into account the regions' revenue collection costs, that is, the center maximizes

$$U_c = \log[(1 - x_1)Y_1 + G_1 - a_1Y_1^2] + \log[(1 - x_2)Y_2 + G_2 - a_2Y_2^2].$$

5. The case where transfer depends on both regions' lobbying efforts is discussed in the fifth section. The assumption that transfer is an increasing function of own lobbying effort is employed by almost all studies on interest group lobbying. As claimed by Findlay and Wellisz (1984), 'the economists' focus is on how the relevant inputs (or lobbying efforts) are connected with ultimate outputs, the details of the process of conversion in each case being inside a black box'.
6. While a firm is owned by the state, the government is responsible for guaranteeing workers' employment and welfare. If the firm is operating at a loss, the government is obliged to offer subsidies. Privatization of state-owned enterprises will, at least theoretically, free the government from being obliged to rescue loss-making enterprises.

NOTES ON CHAPTER 4: INTERGOVERNMENTAL RELATIONS AND MONETARY MANAGEMENT

1. Part of this section draws from Ma and Walker (1993c).
2. Recently, foreign exchange businesses were allowed in many other specialized banks, such as the Agricultural Bank, the Industrial and Commercial Bank, and the Construction Bank.
3. State Statistical Bureau (1994).
4. See Yi (1992) for a detailed description of this process.
5. See World Bank (1990b) for a detailed description of China's money supply fluctuation during 1980–89.
6. Similar practices are used by enterprises to induce (force) the local banks to extend additional loans. Many enterprises divert loans designated for working capital to fixed investment. After these investments are made, they come to the bank to request additional loans for working capital. The shortage of working capital may immediately endanger the continuation of the firms' operation, and the bank has no choice but to give more loans. In other cases, many firms divert loans reserved for wage payments for long-term investment, and then ask the bank for additional loans for wage payments. Facing these pressures, the government, which cares about 'stability', is forced to make additional loans. Through the local governments, these demands from the enterprises are translated into pressures on the central bank.
7. These measures include restabilizing the yuan by selling US dollars in the swap market and raising bank interest rates on deposits, which fell far below the stock and bond returns and the black market rate.
8. In October 1993, the government announced that it was re-establishing control over company payrolls, apparently in an effort to reduce wage increases. According to the announcement, all enterprises, whether state owned or not,

must submit their payroll budgets to local government labor departments for approval. The enterprises could decide for themselves how to divide the budgeted amount among the workers, though. This regulation reversed prior government policy which allowed companies to make business decisions on their own. In July 1992, the government gave state-owned companies the right to set their employees' wages instead of following a government wage scale.

NOTES ON CHAPTER 5: CENTRAL GOVERNMENT CREDIBILITY AND MONETARY EXPANSION: A GAME THEORETIC MODEL

1. Data are from State Statistical Bureau (1993b), (1994). Government borrowings from the central bank are 'hard' deficits in Chinese definition (See State Statistical Bureau 1994, pp. 213–14).
2. Defined as profits plus taxes divided by total sales. See State Statistical Bureau (1993b).
3. The most obvious reason is the painful experience of hyperinflation during a rapid price reform in the transitional economies. Detailed arguments favoring a gradual approach to price reform approach can be found in a large number of papers, see Alexeev (1991), Wiemer (1992), Helpman (1988), and Dewatripont and Roland (1991). The arguments include (1) supplies often respond to the price signals with a significant lag and this supply rigidity will lead to a high inflation in the short-run after price liberalization, (2) in many former socialist countries, the market structure is highly concentrated – the monopolistic behavior under this market structure will create a high inflation and a production disruption following price liberalization, (3) price decontrol will be likely to provoke resistance from the poor stratum of the population as well as the beneficiaries of the old system – the tolerance limit of income redistribution bounds the speed of price liberalization, and (4) a politically weak government, whose primary objective is to stay in power, is likely to be unwilling to face the risk of rapid price reform. Supporting evidence for most of these arguments can be found in the case of China.
4. See Barro and Gordon (1983) and Canzoneri (1985).
5. The market value is not the central government's criterion for selecting projects. Instead, the principle of state investment is to 'fill the gaps' (that is, solve the problem of shortage). Since the early 1980s, the center has consistently called for more investment in bottle-neck sectors, outputs of which have high shadow prices but low market prices. Estimations of shadow prices for the purpose of providing guidelines to state investment have been undertaken by the Development Research Center of the State Council and other institutions. See Ma (1989).
6. Bernanke and Gertler (1987) used a production function with credit as the input. For discussions on credit-output link, see Lown (1988) and Bernanke (1993). In China, credit from the banking system is a main source of fixed assets investment and working capital. Chan *et al.* (1992) studied China's industrial growth and found that total GNP is highly responsive to the input of credit.

7. To see the reason, look at the simplest bank's balance sheet:

A Simplified Bank's Balance Sheet

Assets	Liability
Cash in bank	Deposits
Loan	

Suppose the bank's total liability consists of only deposits and the assets are cash in the bank and loans. The balance sheet states cash in bank + loans = deposits. Define the money supply, M, as the total cash in circulation plus deposits. Therefore

$$\begin{aligned} M &= \text{cash in circulation} + \text{deposits} \\ &= \text{cash in circulation} + (\text{cash in the bank} + \text{loans}) \\ &= \text{total currency} + \text{loans} \end{aligned}$$

If total currency is small relative to outstanding loans in the economy, we have $M \approx L$.

8. Alternatively, it could be assumed that there are a large number of regions, each of which also dislikes inflation. Let the regions be indexed by $i = 1, 2, \ldots n$. The general specification of ith region's objective function is

$$\begin{aligned} &\text{Max } V_i = f(L_{i1}) + \delta f(L_{i2}) - \pi(L_{i1} + L_{-i1}, L_{i2} + L_{-i2}) \\ &\{L_{i1}, L_{i2}\} \\ &\qquad \text{s.t. } L_i = L_{i1} + L_{i2} \end{aligned}$$

where L_i is the credit ceiling for region i, L_{i1} is the credit extended to project 1 in region i and L_{-i1} is the credit to the project in all other regions. Similar definitions are for L_{i2} and L_{-i2}. $\pi(L_{i1} + L_{-i1}, L_{i2} + L_{-i2})$ is the national inflation rate, depending on the total credit extended by all n regions in the economy. We can show that as n becomes large, region i perceives almost no impact of its individual choice of L_{i1} and L_{i2} on national inflation, that is, $\partial\pi/\partial L_{i1}$ is close to zero for a large n. For simplicity we drop the inflation term in the region's objective function in the text.

9. Equation (5.5) does not necessarily mean that the output prices for projects 1 and 2 are 1 and δ, respectively. δ should be viewed as the relative price of project 2's output in terms of project 1's output.

10. Examples of project 2 include those in the agricultural, infrastructure, and energy sectors. Since the early 1980s, the center has repeatedly called for more investment in the agricultural, infrastructure, and energy sectors, but these sectors have not been given priority by the regions. See the State Council (1989).

11. It is obvious that if the center can centralize credit allocation, that is allocate the credit by itself, the two projects will be equally financed.

12. Suppose $\alpha < 1/2$. The $dU/dL > 0$ for $L < L^c$ and $dU/dL < 0$ for $L > L^c$ so L^c is a global maximum. When $\alpha > 1/2$ the inequalities run the wrong way so that it is possible to increase the centers utility without bound by increasing L without bound.
13. This two-stage game is designed to mimic the practice adopted by the Chinese banking system: at the beginning of each year, a credit plan containing credit ceilings on local banks is announced by the central bank; during the implementation, this credit plan is repeatedly revised. A reduced form of this game is a one-stage Cournot game in which the region chooses the credit going to project 1, and the center chooses the credit going to project 2. It yields exactly the same equilibrium as the game described in this paper. However, the version adopted in this paper is able to describe the process of credit ceiling revision and to show why the optimal credit ceiling in the commitment case is not credible – the most important result of this paper.
14. Readers may ask whether the region wants/or is able to divert part of aL to project 1 under the decentralized system without monitoring. The answer is that under the system specified in this section, the region is free to do so. Note that β' is not restricted to values below 1. By choosing a β' greater than 1, the region can not only invest all the initial credit, L, in project 1, but also divert $(\beta' - 1)L$ from the additional credit, aL, to project 1.
15. This recommendation stems from examining the role of assumption 3 in the second section.
16. This recommendation stems from examining the role of assumption 4 in the second section.
17. This recommendation stems from examining the role of assumption 2 in the second section.
18. This recommendation stems from examining the role of assumption 1 in the second section.
19. This set-up is similar to the US Federal Reserve System under which 12 Regional Federal Reserve Banks do not belong to any states. Instead, each serves a number of states.

NOTES ON CHAPTER 6: INTERGOVERNMENTAL RELATIONS AND MARKET DEVELOPMENT

1. See State Statistical Bureau (1992). The rest include foreign-invested enterprises, privately-owned enterprises, and jointly-owned enterprises.
2. Except for a number of minor central-fixed taxes that must be submitted to the center exclusively. See Chapter 1 for a description of central-fixed taxes.
3. See 'State Planning Commission Changes its Function', *People's Daily* (Overseas Edition), 10 May 1994.
4. Interview with enterprise managers in 1993.
5. See Ma (1993a).
6. See Chapter 1.
7. See *People's Daily* (Overseas Edition), 6 May 1994.
8. On 1 May 1986, on the stretch of road linking Sheng and Shaoxing counties, several factories had despatched people out onto the side of the road to open

their sacks, lay out bamboo hoarding, and jack up their purchasing prices. On the other side of the road, 20 or so cadres from the government departments in Sheng county formed up a human chain. Along the road came 400–500 farmers with their tea. When they arrived at the county line the farmers observed that they were being blocked in front, and from behind people were calling out 'whoever's price is high, come and sell over here!' (Forster, 1991)

9. Watson and Findlay (1992).
10. Zhang *et al.* (1991).
11. The following few paragraphs are based on Watson and Findlay (1992).
12. Prior to the early 1980s, the state was the only purchaser of most agricultural and raw material products. The situation that triggered the 'commodity wars' was the abolition of the unified purchasing system during 1984–5. In the case of cotton, in 1985, the old centralized marketing and price system was replaced by one in which the government purchased through contracts. After meeting certain quota, cotton producers were allowed to sell at higher prices for above quota output or at the free market. The case was similar with tea, as its procurement was monopolized by the central authority before 1984. The central government relaxed control in 1984, allowing various trading units to procure tea at negotiable prices. Provincial trading companies were also permitted to export tea directly, as opposed to the previous case when only the central Tea Import and Export Company was designated to trade internationally. Although these reforms intended that the state contracts would be filled before the products entered the free market, commercial departments, processing enterprises and private dealers all competed for supplies.
13. See Agarwala (1992).
14. Interview with a researcher at the Development Research Center of the State Council, China, 1993.
15. See 'Inter-provincial Market Established', *Beijing Review*, 27 December 1993–2 January 1994.
16. See Pi and Pan (1995).
17. The State Council's 'Interim Regulation of Promoting Economic Cooperation', 1980. See Gu *et al.* (1992).
18. Ibid.
19. Ibid.
20. Ibid.
21. *Law Daily* (Fazhi Ribao), 11 November 1993, presents such a case.
22. In Xiangtan City (Hunan Province), a firm which was forced to join an enterprise group declared to 'withdraw' from the group. See *Economic Daily* (Jin Ji Ri Bao), 8 May 1993.
23. See 'A State Council Leader Says "Changing Name" Companies Is Not a Reform, but a Retreat from Reform', *Guangming Daily*, 2 March 1993.
24. For example, the 'No.2 Automobile Group' was designated to submit less than 30 per cent of its profits according to its contract with the government. The Capital Steel Company's profit retention rate is as high as 80 per cent. On average, large and medium state enterprises retain only approximately 20 per cent of their profits. A comparison made between the Capital Steel Company (with an 80 per cent profit retention rate) and the Shanghai Metal Bureau (with a profit retention rate less than 20 per cent) indicates that the Capital Steel

Company's performance (growth of output and profits) critically depended on the formula of profit division between the state and the enterprise. See Ma (1993b).

25. See Wu and Li (1993).
26. See Sender (1993).
27. Interview with a lawyer with Reichler and Soble at Washington, DC.
28. See Wu (1993).
29. See 'The State Council's Decision Concerning the Main Points of Current Industrial Policies', *People's Daily* (Overseas Edition), 18 March 1989.
30. A line ministry is a specialized ministry in charge of an industrial sector.
31. One famous case under the Commerce Clause, Gibbon vs Ogden (1824), concerned the New York state legislature's grant of exclusive rights to Livingston and Fulton to operate steamboats in New York waters. This grant of monopoly clearly restricted open market competition. Brown vs Maryland (1827) invalidated a $50 license fee imposed by the state for the sale of imported goods. State power of this sort would have allowed significant regulation of the economy to local advantage in the form of non-tariff barriers. See Weingast (1993).
32. See World Bank (1994b).
33. See Lee *et al.* (1992).
34. China's National People's Congress recently passed the Anti-unfair Competition Law. With only a brief mention of the prohibition of abuses of administrative powers, the law emphasizes the regulations on conduct of individual enterprises. The law is particularly concerned with violations of patent law and trademark law, false or misleading advertisements, lottery sales, forced underselling, cartelling, and molesting of competitors' businesses.

Bibliography

Agarwala, R. (1992) *China: Reforming Intergovernmental Fiscal Relations*, Discussion Paper no. 178 (Washington, DC: World Bank).

Alesina, A. and Summers, L. (1993) 'Central Bank Independence and Macroeconomic Performance: Some Comparative Evidence', *Journal of Money, Credit, and Banking*, vol. 25, no. 2, pp. 151–62.

Alexeev, M. (1991) 'If Market Clearing Prices Are So Good then Why Does not (Almost) Everybody Want Them?', *Journal of Comparative Economics*, vol. 15, pp. 380–90.

An, T., Zhou, S. and Zhu, Y. (1991) *Overview of Fiscal and Monetary (Polices) (Caizhen Jinrong Gailun)*, (Beijing: China Financial Economic Press).

Bahl, R. and Wallich, C. (1992) *Intergovernmental Fiscal Relations in China*, Policy Research Working Paper no. 863 (Washington, DC: World Bank).

Barro, R. and Gordon, D. (1983) 'Rules, Discretion and Reputation in a Model of Monetary Policy', *Journal of Monetary Economics*, vol. 12, pp. 101–22.

Becker, G. S. (1983) 'A Theory of Competition Among Pressure Groups for Political Influence', *The Quarterly Journal of Economics*, vol. 18, no. 3, pp. 371–400.

Bell, M., Khor, H. E., Kochhar, K., Ma, J., N'guiamba, S. and Lall, R. (1993) *China at the Threshold of a Market Economy*, Occasional Paper, no. 107 (Washington, DC: International Monetary Fund).

Bernanke, B. S. (1993) 'Credit in the Macroeconomy', *Federal Reserve Bank of New York Quarterly Review*, vol. 18, pp. 50–70.

Bernanke, B. S. and Gertler, M. (1987) 'Banking and Macroeconomic Equilibrium', in W. Barnett and K. Singleton (eds), *New Approaches to Monetary Economics* (Cambridge: Cambridge University Press).

Blejer, M. I. (1993) 'China: Prolonged Reforms and the Weakening of Fiscal Control', in V. Tanzi (ed.), *Transition to Market: Studies in Fiscal Reform* (Washington, DC: International Monetary Fund).

Bo, D. X. and Li, M. S. (1991) 'Basic Strategies to Address the Current Fiscal Difficulties', (Zouchu Muqian Caizhen Qunjing de Jiben Duicesilu), *Economic Research* (Jingji Yanjiu), no. 12, pp. 53–8.

Bomfin, A. N. and Shah, A. (1991) *Macroeconomic Management and the Division of Powers in Brazil: Perspectives for the Nineties*, Policy Research Working Paper no. 567 (Washington, DC: World Bank).

Canzoneri, M. (1985) 'Monetary Policy Games and the Role of Private Information', *American Economic Review*, vol. 75, no. 5, pp. 1056–170.

Cardoso, E. and Yusuf, S. (1994) 'Red Capitalism: Growth and Inflation in China', *Challenge*, May–June, pp. 49–56.

Chan, M. W., Deaves, R., Luke, R. D. and Cheng, W. (1992) 'An Analysis of Money and Output in the Industrial Sector in China', *Journal of Asian Economy*, vol. 3, no. 2, pp. 271–80.

Chow, G. C. (1987) 'Money and Price Level Determination in China', *Journal of Comparative Economics*, vol. XI, pp. 319–33.

Clarke, D. C. (1992) 'Regulation and its Discontents: Understanding Economic Law in China', *Stanford Journal of International Law*, vol. 28, no. 2, pp. 283–322.

Corcoran, W. J. (1984) 'Long-Run Equilibrium and Total Expenditures in Rent-Seeking', *Public Choice*, vol. 43, pp. 89–94.

Development Research Center of the State Council (1993) *Almanac of China's Economy 1993 (Zhongguo Jingji Nianjian 1993)* (Beijing: Economics and Management Press).

Development Research Center of the State Council (1994) *Almanac of China's Economy 1994 (Zhongguo Jingji Nianjian 1994)* (Beijing: Economics and Management Press).

Dewatripont, M. and Roland, G. (1991) *'The Virtues of Gradualism and Legitimacy in the Transition to a Market Economy'*, Center for Economic Policy Research Discussion Paper, no. 538 (London).

Economic Research Center of the State Council (1981) *Almanac of China's Economy 1980 (Zhongguo Jingji Nianjian 1980)* (Beijing: Economics and Management Press).

Economic System Reform Commission (1992a) 'Proposed Norms of Limited Liability Companies' (Youxian Zeren Gongsi Guifan Yijian), *People's Daily* (Overseas Edition), June 20.

Economic System Reform Commission (1992b) 'Standards for Limited Liability Stock Companies' (Gufen Gongsi Guifan Yijian), *People's Daily* (Overseas Edition), June 20.

Editorial Department of China Finance (1995) 'Our Country's Finance is Developing in a Stable Environment' (Woguo Jinrong Zai Wending Zhong Jiankang Fazhan), *China Finance (Zhongguo Jinrong)*, no. 417, p. 1.

Fan, G. (1989) 'Reform, Adjustment, Growth, and Friction Inflation', *Economic Research (Jingji Yanjiu)*, no. 1, pp. 13–25.

Fan, G., Zhang, S. and Wang, L. (1993) 'Dual-track Transition and Dual Track Regulation and Control – a Study of the Characteristics of Economic Fluctuations in China Since the Beginning of the Reforms (Part I)', *Economic Research (Jingji Yanjiu)*, no. 10, pp. 15–26.

FBIS (1993) 'Bank Official Views Upcoming Monetary Reform', *Daily Report China*, 13 December 1993.

Feltenstein, A. and Ziba, F. (1987) 'Fiscal Policy, Monetary Target, and the Price Level Determination in a Centrally Planned Economy: an Application to the Case of China', *Journal of Money, Credit, and Banking*, vol. 19, no. 2, pp. 137–56.

Feltenstein, A. and Ha, J. (1989) *Measurement of Repressed Inflation in China: the Lack of Coordination Between Monetary Policy and Price Control*, Working Paper (Kansas: Department of Economics, University of Kansas).

Findlay, C., Will, M. and Watson, A. (1992) *Policy Reform, Economic Growth and China's Agriculture* (Washington, DC: World Bank, Mimeo).

Findlay, R. and Wellisz, S. (1984) 'Toward A Model of Endogenous Rent-Seeking', in D. C. Colander (ed.), *Neoclassical Political Economy: The Analysis of Rent-Seeking and DUP Activities* (Cambridge, Mass.: Ballinger Publishing Company).

Forster, K. (1991) *China's Tea War* (Chinese Economy Research Unit, Adelaide, Australia: University of Adelaide, Mimeo).

Fudenberg, D. and Tirole, J. (1991) *Game Theory* (Cambridge, Mass.: The MIT Press).

Fujiwara, T. (1992) *Intergovernmental Fiscal Relations in Japan: Lessons from Postwar Reform and Evolution Thereafter* (Washington, DC: Africa II Department, Mimeo, World Bank).

Garnaut, R. and Ma, G. (1993) 'Economic Growth and Stability in China', *Journal of Asian Economics*, vol. 4, pp. 5–24.

Gordon, R. (1983) 'Optimal Taxation Approach to Fiscal Federalism', *Quarterly Journal of Economics*, vol. 98, pp. 567–86.

Gordon, R. (1990) *Macroeconomics*, 5th edn (Glenview, IL: Scott'Foresman-Little).

Gu, S., Lu, P., Hu, J. and Xu, C. (1992) 'The History, *Status Quo*, and Recommendations of Enterprise Groups in China', *Enterprise Group Studies* (Qiye Jituan Yanjiu), vol. VI.

Helpman, E. (1988) 'Macroeconomic Effects of Price Controls: the Role of Market Structure', *The Economic Journal*, vol. 98, pp. 340–54.

Hoyt, W. (1992) *Pre-commitment and State and Local Policy Coordination*, Working Paper (Washington, DC: Georgetown University Economics Department).

Hu, A. and Wang, S. (1994) *China: Report of Government Capacity* (*Zhongguo Guojia Nengli Baogao*) (Liaoning Province: Liaoning People's Press).

Huang, Y. (1990) 'Web of Interests and Patterns of Behavior of Chinese Local Economic Bureaucracies and Enterprises During Reforms', *The China Quarterly*, vol. 123, pp. 431–58.

International Monetary Fund (IMF) (1994) *Government Finance Statistics* (Washington, DC: IMF).

Jia, H. and Lin, Z. (1994) *Changing Central–Local Relations in China* (Boulder, CO: Westview Press).

Ishihara, N. (1993) 'The Local Public Finance System', in Shibata, T. (ed.), *Japan's Public Sector: How the Government is Financed* (Tokyo: University of Tokyo Press), pp. 141–66.

Kydland, F. and Prescott, E. (1977) 'Rules Rather Than Discretion: the Inconsistency of Optimal Plans', *Journal of Political Economy*, vol. 85, pp. 473–90.

Labor Department and Economic System Reform Commission (1992) 'Tentative Regulations of Labor and Wages in Trial Stock Companies' (Gufenzhi Shidian Qiye Laodong Gongzi Shixing Banfa), *People's Daily* (Overseas Edition), 13 July.

Lee, K. U., Urata, S. and Choi, I. (1992) 'Industrial Organization: Issues and Recent Developments', in V. Corbo and S. M. Suh (eds), *Structural Adjustment in a Newly Industrialized Country* (Baltimore, MD: The Johns Hopkins University Press).

Li, J. and Jiang, J. (1993) 'Li Lanqing Addresses Grain Procurement Meeting', *Daily Report: China*, Federal Broadcast Information Services, 28 May, pp. 35–6.

Li, Y. (1992) 'Local Forces Will Dominate Economic Reform in China', *Central Daily* (*Zhong Yang Ri Bao*), 23 October.

Liu, G., Li, J., Qiu, X., Liu, S., Wang, T., Zhang, Y. and He, J. (1994) 'An Analysis and Forecast of China's Economic Situation', *Economic Research* (*Jingji Yanjiu*), no. 5., pp. 3–10.

Lown, C. S. (1988) 'The Credit-output Link vs. the Money-output Link: New Evidence', *Economic Review: Federal Reserve Bank of Dallas*, November, pp. 1–9.

Ma, J. 1989 'Problems in Industrial Structural Change and the Direction of Reform', *Economic Theory and Economic Management (Jinji Lilun Yu Jinji Guanli)*, no. 1.

Ma, J. (1993a) *China: Fiscal Policy Differentials Across Regions* (Washington, DC, Mimeo, World Bank).

Ma, J. (1993b) *Reforming Corporate Governance Structure in China* (Washington, DC, Mimeo, World Bank).

Ma, J. (1994a) 'Intergovernmental Fiscal Relations in Japan and Korea', paper prepared for the Senior Policy Seminar on Intergovernmental Fiscal Relations in China (12–17 September 1994, Dalian, China) jointly sponsored by Dongbei University of Finance and Economics (DUFE), Economic Development Institute of the World Bank, and the Ministry of Finance, China. Also appear in Economic Development Institute Working Paper (No. 94–41), World Bank.

Ma, J. (1995a) 'Modelling Central–Local Fiscal Relations in China', *China Economic Review*, vol. 6, no. 1, pp. 105–37.

Ma, J. (1995b) *Macroeconomic Management and Intergovernmental Relations in China*, Policy Research Working Paper Series no. 1408 (Washington, DC: the World Bank).

Ma, J. (1995c) 'Defining the Limits of Local Government Power in China', *The Journal of Contemporary China*, no. 10 (Fall), pp. 3–22.

Ma, J. (1995d) 'China: Central Government Credibility and Economic Overheating', *Economic Systems*, vol. 19, no. 3 (September), pp. 237–61.

Ma, J. (1995e) 'The Reform of China's Intergovernmental Fiscal Relations', *Asian Economic Journal*, vol. 9, no. 3 (November), pp. 205–32.

Ma, J. (1996a) 'Monetary Management and Intergovernmental Relations in China', *World Development*, vol. 24, no. 1 (January), pp. 145–53.

Ma, J. (1996b) 'China's Banking Sector: From Liberalization to a Regulatory Framework', forthcoming in *Journal of Contemporary China*.

Ma, J. (1996c) 'Intergovernmental Fiscal Transfer: a comparison of Nine Large Countries', Mimeo, Economic Development Institute (Washington, DC: World Bank).

Ma, J. and Zou, G. (1991) 'Regional Comparative Advantage of Industrial Development in China', *Management World (Guan Li Shi Jie)*, no. 4, pp. 138–46.

Ma, J. and Bell, M. (1992) *China: Regional Policy and Its Macroeconomic Implications*, (Washington, DC, Mimeo, International Monetary Fund).

Ma, J. and Walker, D. (1993) *China's Financial Development in a Reform Era*, (Washington, DC: Georgetown University, Mimeo).

Ma, J. and Li, Y. (1994) 'China's Regional Economic Policy: Effects and Alternatives', *Asian Economic Journal*, vol. 8, no. 1, pp. 39–58.

Ma, J. L. and Luo, L. (1994) 'Important Reforms in China's Tax System', *(JETRO) China Newsletter (Japan)*, no. 110, pp. 12--17.

McKinnon, R. I. (1993) *Gradual versus Rapid Liberalization in Socialist Economies: Financial Policies and Macroeconomic Stability in China and Russia Compared*, paper for the Annual Bank Conference on Development Economics (Washington, DC: World Bank).

Metzger, M. B., Mallor, J. P., Barnes, A. J., Bowers, T. and Phillips, M. J. (1987) *Business Law and the Regulatory Environment: Concepts and Cases* (Homewood, IL: Irwin, Inc.).

Ministry of Finance (1988) *China Fiscal Statistics 1952–1987* (Beijing: China Fiscal Press).

Ministry of Finance (1992) *China Fiscal Statistics 1952–1991* (Beijing: China Fiscal Press).

Mueller, D. C. and Murrell, P. (1986) 'Interest Groups and the Size of Government', *Public Choice*, vol. 48, pp. 125–45.

Mueller, D. C. (1989) *Public Choice II* (Cambridge: Cambridge University Press).

Musgrave, R. (1965) *Essays in Fiscal Federalism* (Washington, DC: Brookings Institution).

Musgrave, R. (1983) 'Who Should Tax, Where and What?' in C. Mclure, Jr (ed.), *Tax Assignment in Federal Countries* (Canberra: Australian National University Press).

Oates, W. E. (1972) *Fiscal Federalism* (New York: Harcourt Brace Jovanovich, Inc.).

O'Brien, K. J. (1992) 'Bargaining Success of Chinese Factories', *The China Quarterly*, vol. 132, pp. 1086–100.

Oksenberg, M. and Tong, J. (1991) 'The Evolution of Central–Provincial Fiscal Relations in China, 1971–1984: the Formal System', *The China Quarterly*, vol. 125, pp. 1–32.

Pechman, J. A. (1985) *Federal Tax Policy*, 5th edn (Washington, DC: the Brookings Institute).

People's Bank of China (1993) *People's Bank of China Annual Report 1992* (Beijing: People's Bank of China).

Peres-Castrillo, D. J. (1992) 'A General Analysis of Rent-Seeking Games', *Public Choice*, vol. 773, pp. 335–50.

Persson, M., Persson, T. and Svensson, L. E. O. (1987) 'Time Consistency of Fiscal and Monetary Policy', *Econometrica*, vol. 55, pp. 1419–32.

Persson, T. and Tabellini, G. (1990) *Macroeconomic Policy, Credibility and Politics* (Chur, Switzerland: Harwood Academic Publishers).

Persson, T. and Tabellini, G. (1993) *Federal Fiscal Constitutions: Part I: Risk Sharing and Moral Hazard*, Mimeo (Stockholm, Sweden: University of Stockholm, Institute for International Studies).

Pi, S. and Pan, G. (1995) 'Illegal Market Protection to Restrict Competition' (Fengsuo Shichang Xianzhijinzheng: Weifa), *Economic Daily* (*Jingji Ri Bao*), 6 January.

Prachowny, M. (1994) *The Goals of Macroeconomic Policy* (London and New York: Routledge).

Qian, Y. (1993) *Japan's Main Bank System and the Relevance for China's Banking Reform*, (Economics Department, Mimeo, Stanford CA: Stanford University).

Rogers, C. (1987) 'Expenditure Taxes, Income Taxes, and Time-Inconsistency', *Journal of Public Economics*, vol. 32, pp. 215–30.

Rogoff, K. (1985) 'The Optimal Degree of Commitment to an Intermediate Monetary Target', *Quarterly Journal of Economics*, vol. 100, pp. 1169–90.

Sender, H. (1993) 'Party of the First Part', *Far Eastern Economic Review*, 11 February, pp. 42–4.

Shah, A. (1991a) *Perspectives on the Design of Intergovernmental Fiscal Relations*, Working Paper WPS 726, Policy Research Department, (Washington, DC: World Bank).

Shah, A. (1991b) *The New Fiscal Federalism in Brazil*, World Bank Discussion Paper 124, Policy Research Department (Washington, DC: World Bank).

Shah, A. (1994) *The Reform of Intergovernmental Fiscal Relations in Developing and Emerging Market Economies*, Policy and Research Series no. 23 (Washington, DC: World Bank).

Shen, L. and Dai, Y. (1990) 'On the Formation, Effects and Causes of China's Lords Economy', *Economic Research (Jinji Yanjiu)*, no. 3, pp. 12–20.

Sicular, T. (1990) *Plan, Market and Inflation: Potential Problems with China's Two-Track System*, Discussion Paper no. 1501 (Harvard Institute of Economic Research, Boston, MA: Harvard University).

Singh, I. (1992) *China: Industrial Policies for an Economy in Transition*, World Bank Discussion Paper no. 143 (Washington, DC: World Bank).

State Council (1981) 'The State Council's Circular on the Introduction of a Public Finance Management System by Which the Scope of Revenue and Expenditure will be Divided for Central and Local Governments Respectively and Each Will be Responsible for the Balance of Its Own Budget', in Economic Research Center for the State Council (ed.) *Almanac of China's Economy 1981* (Beijing: Economics and Management Press).

State Council (1986) 'Regulations on the Public Finance Management System of Dividing Taxes, Defining Revenues and Expenditures, and Fiscal Contracts at Each Levels of Government', in Economic, Technological, and Social Development Research Center for the State Council (ed.) *Almanac of China's Economy 1986* (Beijing: Economics and Management Press).

State Council (1989) 'Decisions Concerning the Main Points of Industrial Policies', *People's Daily* (Overseas Edition), 18 March.

State Council (1992) 'Rules of Transforming Management of State Enterprises', *People's Daily* (Overseas Edition), 24 July.

State Council (1993) *State Council's Decision Regarding the Implementation of Tax Assignment Fiscal Management System* (Beijing: State Council, 1993).

State Property Management Bureau, Ministry of Finance and State Industrial and Commerce Administration Bureau (1992) 'Trial Measures of State Assets Registration', *People's Daily* (Overseas Edition), 5 July.

State Property Management Bureau, Economic System Reform Committee (1992) 'Provisional Regulations on State Assets Management in Enterprises Experimenting with Joint Stock System', *People's Daily* (Overseas Edition), 27 July.

State Statistical Bureau (1989) *Forty Years of China's Development (Zhongguo Fenjin Sishi Nian)* (Beijing: China State Statistical Press).

State Statistical Bureau (1991a) *China's Finance Yearbook 1990 (Zhongguo Jingrong Ningjian 1990)* (Beijing: China State Statistical Press).

State Statistical Bureau (1991b) *China's Statistical Yearbook 1991* (Beijing: China State Statistical Press).

State Statistical Bureau (1992a) *China's Statistical Yearbook 1992* (Beijing: China State Statistical Press).

State Statistical Bureau (1992b) *China's Industrial Statistical Yearbook 1991* (Beijing: China State Statistical Press).

State Statistical Bureau (1993a) *China's Industrial Statistical Yearbook 1993* (Beijing: State Statistical Press).

State Statistical Bureau (1993b) *China's Statistical Yearbook 1993* (Beijing: China State Statistical Press).

State Statistical Bureau (1994) *China's Statistical Yearbook 1994* (Beijing: China State Statistical Press).

State Statistical Bureau (1995) 'Statistical Report of National Economy and Social Development in 1994 by the State Statistical Bureau of the People's Republic of China', *People's Daily* (Overseas Edition), 2 March, p. 2.

Tsang, S. K. and Cheng, Y. S. (1994) 'China's Tax Reform of 1994', *Asian Survey*, vol. 34, no. 9, pp. 769–88.

Tullock, G. (1980) 'Efficiency Rent-Seeking', in J. M. Buchanan, R. D. Tollison and G. Tullock (eds), *Towards a Theory of the Rent-Seeking Society* (College Station: Texas A & M Press).

van Winden, V. and Frans, A. A. M. (1988) 'The Economic Theory of Political Decision-making', in Julien V. D. Broeck (ed.), *Public Choice* (Boston, Mass: Kluwer Academic Publishers).

Walker, D. (1989) 'Effects of Economic Reforms on the Banking System in China', *Economia International*, vol. 42, pp. 244–57.

Wang, S. (1994a) 'Central–Local Fiscal Politics in China', in H. Jia and Z. Lin (eds), *Changing Central–Local Relations in China* (Boulder, NY: Westview Press).

Wang, S. (1994b) *The Limit of Decentralization*, Paper presented in the Conference on Rule Based Power Division and System Reform in China, sponsored by the Center for Modern China, Princeton University, 21–23 October.

Wang, Y. (1991) 'Government Reform, Fixed Capital Investment Expansion, and Inflation: a Behavior Model Based on the Chinese Experience', *China Economic Review*, vol. 2, no. 1, pp. 3–27.

Watson, A. and Findlay, C. (1992) 'The "Wool War" in China', in C. Findlay (ed.), *Challenges of Economic Reform and Industrial Growth: China's Wool War* (Australia: Allen & Unwin).

Weingast, B. R. (1993) *The Political Foundations of Antebellum American Economic Growth* (San Francisco, CA: Hoover Institution, Stanford University).

Welch, S., Gruhl, J., Steinman, M., Comer, J. and Basehart, H. (1993) *Understanding American Government*, 2nd edn (St Paul, MN: West Publishing Company).

Wiemer, C. (1992) 'Price Reform and Structural Change: Distributional Impediments to Allocative Gains', *Modern China*, vol. 18, no. 2, pp. 171–96.

Wong, C. (1991) 'Central–Local Relations in an Era of Fiscal Decline: the Paradox of Fiscal Decentralization in Post-Mao China', *The China Quarterly*, vol. 128, pp. 691–715.

Woo, W. T., Fan, G., Hai, W. and Jin, Y. (1993) 'The Efficiency and Macroeconomic Consequences of Chinese Enterprise Reform', *China Economic Review*, vol. 4, no. 2, pp. 153–68.

World Bank (1989) *World Development Report 1989: Financial System and Development* (Oxford: Oxford University Press, 1989).

World Bank (1990a) *China: Revenue Mobilization and Tax Policy* (Washington, DC: World Bank).

World Bank (1990b) *China: Financial Sector Policies and Institutional Development* (Washington, DC: World Bank).

World Bank (1991) *China: Macroeconomic Stability and Planning Under a Decentralized System* (Washington, DC: World Bank).

World Bank (1993) *China: Budgetary Policy and Inter-governmental Relations* (Washington, DC: World Bank, Mimeo).

World Bank (1994a) *China: Macroeconomic Stability in a Decentralized Economy* (Washington, DC: World Bank,).

World Bank (1994b) *China: Internal Market Development and Regulations* (Washington, DC: World Bank).

Wu, B. and Li, Y. (1993) 'The Truck Incident', *Law Daily* (*Fazhi Ribao*), 11 December.

Wu, M. (1990) 'Thoughts on Local Government Behavior', *Economic Research* (*Jinji Yanjiu*) no. 7, pp. 56–60.

Wu, N. (1993) 'Legal Service For Economic Order', *Beijing Review*, vol. 34, no. 32, 9–15 August, p. 4.

Yi, G. (1992) 'The Money Supply Mechanism and Monetary Policy In China', *Journal of Asian Economics*, vol. 3, no. 2, pp. 217–38.

Yuan, D. (1992) 'Problems of and Solutions to the Central–Local Fiscal Relations', *China Economic Problems* (*Zhongguo Jingji Wenti*), no. 1, pp. 10–13.

Zhang, J. (1995), 'Tax Assignment System and Fiscal Reform', in Finance Yearbook Editorial Committee ed., *Finance Yearbook of China 1995 (Zhongguo Caizheng Nianjian),* (Beijing: China Finance Journal), pp. 61–3.

Zhang, X., Lu, W., Sun, K., Findlay, C. and Watson, A. (1991) *The 'Wool War' and the 'Cotton Chaos': Fibre Marketing in China*, Working Paper no. 14 Chinese Economy Research Unit (Adelaide, Australia: University of Adelaide).

Zhu, J. and Wu, M. (1992) 'Management of Central and Local Government: Barriers and Choices', *Economic Research* (*Jingji Yanjiu*), no. 8, pp. 15–21.

Zhu, M. (1993) *The Mechanics and Consequences of China's Revenue Contract System*, Policy Research Department (Washington, DC: Mimeo, World Bank).

Zou, H. F. (1993), *Optimal Design of Federal Grants to Localities*, Policy Research Department (Washington, DC: Mimeo, World Bank).

Index